20 Springer Series in Solid-State Sciences

Edited by Peter Fulde

W0050126

Springer Series in Solid-State Sciences

Editors: M. Cardona   P. Fulde   K. von Klitzing   H.-J. Queisser

---

Managing Editor: H. K. V. Lotsch

Morikazu Toda

Theory of Nonlinear Lattices

Second Enlarged Edition

With 40 Figures

Springer-Verlag Berlin Heidelberg New York
London Paris Tokyo

Professor *Morikazu Toda*

Professor emeritus, Tokyo University of Education
Home Address: 5-29-8-108 Yoyogi, Shibuya-ku, Tokyo 151, Japan

Series Editors:
Professor Dr., Dr. h. c. Manuel Cardona
Professor Dr., Dr. h. c. Peter Fulde
Professor Dr., Dr. h. c. Klaus von Klitzing
Professor Dr. Hans-Joachim Queisser

Max-Planck-Institut für Festkörperforschung, Heisenbergstrasse 1
D-7000 Stuttgart 80, Fed. Rep. of Germany

Managing Editor:
Dr. Helmut K. V. Lotsch

Springer-Verlag, Tiergartenstrasse 17
D-6900 Heidelberg, Fed. Rep. of Germany

ISBN-13: 978-3-540-18327-3   e-ISBN-13: 978-3-642-83219-2

DOI: 10.1007/978-3-642-83219-2

Library of Congress Cataloging-in-Publication Data. Toda, Morikazu, 1917- Theory of nonlinear lattice. (Springer series in solid-state sciences; 20) Revised translation of: hisenkei koshi rikigaku. Includes bibliography and index. 1. Lattice dynamics. 2. Nonlinear theories. I. Title. II. Series. QC176.8.L3T6213 1988 530.1'5133 88-21838

This work is subject to copyright. All rights are reserved, whether the whole or part of the material is concerned, specifically the rights of translation, reprinting, reuse of illustrations, recitation, broadcasting, reproduction on microfilms or in other ways, and storage in data banks. Duplication of this publication or parts thereof is only permitted under the provisions of the German Copyright Law of September 9, 1965, in its version of June 24, 1985, and a copyright fee must always be paid. Violations fall under the prosecution act of the German Copyright Law.

© Springer-Verlag Berlin Heidelberg 1981 and 1989

Softcover reprint of the hardcover 2nd edition 1989

The use of registered names, trademarks, etc. in this publication does not imply, even in the absence of a specific statement, that such names are exempt from the relevant protective laws and regulations and therefore free for general use.

2154/3150-543210 – Printed on acid-free paper

Preface to the Second Edition

Since the first edition of this book was published in 1981, seven years have passed, and in these years the subject has undergone considerable developments in a number of directions. In this second edition a new chapter has been added to describe several important recent achievements. Since the new developments are so far-reaching, it has become nearly impossible to cover all the areas touched upon by these developments within the framework of the book. Nevertheless the author has tried to elucidate such problems as integrability, generalized lattices, the Bethe ansatz and so forth. Some numerical results in which the author is especially interested are also included.

The opportunity has been taken to correct a few typographical errors from the first edition and some figures have also been corrected.

June 1988   *Morikazu Toda*

Preface to the First Edition

This book deals with waves in lattices composed of particles interacting by nonlinear forces. Since motion in a lattice with exponential interaction between nearest neighbors can be analyzed rigorously, it is treated as the central subject to be discussed.

From the idea that the fundamentals of the mathematical methods for nonlinear lattices would be elucidated by rigorous results, I was led in 1966 to the lattice with exponential interaction, which has since proved to be a subject of intensive investigation by many researchers. Therefore I have tried to describe the development of the study of this lattice. The presentation is intended to be coherent and self-contained.

Chapter 1 starts with a rather historical exposition, and deals with the motion in the lattices and in continuous systems in general. Fundamental concepts necessary for later chapters, including the particle-like behavior of stable pulses (solitons), the most characteristic entities of the nonlinear waves, are introduced. The dual transformation, which exchanges the roles of particles and interaction, is described for development in the next chapter.

Chapter 2 is devoted to the lattice with exponential interaction and shows that it has particular solutions such as periodic waves (cnoidal waves), solitons, and multisoliton solutions. The continuum approximation for the lattice, an equivalent nonlinear LC circuit, and related nonlinear phenomena are discussed. Furthermore, the results presented from the numerical calculations demonstrate that the trajectories in the coordinate-momentum space of this lattice are very smooth. The recognition of this fact led to the discovery of other analytic conserved quantities besides the total momentum and the total energy. The existence of many conserved quantities indicates that the equations of motion for this lattice are integrable.

In Chapter 3 the equations of motion are written in matrix form. First the conserved quantities are derived, followed by a presentation of the method of deriving solutions for an infinite lattice under given initial conditions when the motion is confined to a finite region. For this purpose a discrete wave equation is introduced for a wave scattered by the motion in the lattice; this is the so-called inverse scattering method. Also discussed is another method of solving the inverse problem for a finite lattice using conserved quantities.

In Chapter 4, the initial value problem for a periodic lattice is introduced. Use is made of a discrete wave equation with the periodic potential due to the motion in the lattice, and its spectrum and auxiliary spectrum are used to find the solution of the inverse problem, which has parameters to be determined from the initial data. Since the general solution is so complicated, some concrete problems are presented, which include the derivation of a cnoidal wave from a one-gap spectrum, the calculation of the spectrum for a motion with small amplitude, and a detailed calculation for a three-particle system.

Chapter 5 starts with the three-particle periodic system to integrate the equations of motion following the general principles of mechanics, deriving action and angle variables which characterize the quasi-periodic motion in the lattice. A simple example is also shown.

Appendices have been added following the order in the text, with further ones for general comments.

Since the description is limited to rigorous analytic treatments, interesting but unfinished analytic problems, numerically established phenomena which are not solved analytically, and related problems in solid-state physics and field theory are not included. Topics not discussed are, for instance, wave propagation in a lattice with an impurity or impurities, the possibility of the existence of localized modes, nonlinear lattices in higher dimensions, the reflection of waves at a free end of the lattice, successive fracture phenomena of a modified nonlinear lattice, numerical experiments on the chaotic motion of the general nonlinear lattices, and the ergodic problem. Quantization of the nonlinear lattice is also a future problem.

To write this book I referred to articles and review works I wrote, and to the notes for the courses given at some universities.

This is an English translation of the book originally published in Japanese in 1978. In the present edition some parts have been revised and others added for better understanding.

The subject has undergone considerable development in these years, but mainly in mathematical extensions which have emerged from the content of the text. In this issue bibliographical notes have been added to include this recent research.

It is my great pleasure to acknowledge the understanding and encouragement of Professors H. Wegeland, N. Zabusky, and J. Ford. Thanks are also due my colleagues and friends for their assistance, especially to Professor T. Kotera, and the members of the seminar I have had for many years.

I wish to thank the staff of Springer-Verlag for much valuable assistance during the course of publication. I am also indebted to Miss M. Ohishi for the excellent work of typewriting the manuscript.

January 1981   *Morikazu Toda*

Contents

1. Introduction

Exact treatment of oscillation in nonlinear lattices became serious in the early 1950's, when *Fermi, Pasta*, and *Ulam* (FPU) numerically studied the problem of energy partition. For a linear lattice normal modes of oscillation are mutually independent, and no energy exchange takes place among these modes. Fermi et al. thought that, if nonlinearity of interaction is introduced, energy flow would take place among these linear modes, finally establishing equipartition of energy of statistical mechanics, and wanted to verify this assertion by numerical experiments. However, contrary to their expectation, only a little energy partition occurred, and the state of the systems was found to return periodically to the initial state. In this chapter, we begin with this problem and describe its characteristic features and the starting point of theoretical arguments.

1.1 The Fermi-Pasta-Ulam Problem

The problem we consider in this book is classical mechanics of one-dimensional lattices (chains) of particles with nearest neighbor interaction. Restricting ourselves to a uniform system (without impurities), let m denote the mass of each particle, y_n the displacement of the nth particle, and $\phi(y_{n+1} - y_n)$ the interaction potential (of the spring) between neighboring particles. Then the equation of motion is given by

$$m\frac{d^2y_n}{dt^2} = \phi'(y_{n+1} - y_n) - \phi'(y_n - y_{n-1}) \qquad (n = \cdots, -1, 0, 1, 2, \cdots) \quad (1.1.1)$$

where ϕ' is the derivative of ϕ. Thus

$$f(r) = -\phi'(r) = -d\phi(r)/dr \qquad (1.1.2)$$

is the force of the spring when it is stretched by the amount r, and

$$r_n = y_{n+1} - y_n \qquad (1.1.3)$$

in the mutual displacement.

When $f(r)$ is proportional to r, that is, when Hooke's law is obeyed, the spring is said to be linear, and the potential can be written as

$$\phi(r) = \frac{\kappa}{2} r^2 .$$ (1.1.4)

Then the equations of motion take the form

$$m \frac{d^2 y_n}{dt^2} = \kappa(y_{n+1} - 2y_n + y_{n-1}) .$$ (1.1.5)

Usually y_n is taken as longitudinal displacement. Though transverse wave is more convenient for illustration, nonlinear force is rather hard to be seen when the motion is expressed by transverse displacements. It is more appropriate to think of torsional motion of a system of rods or disks connected by nonlinear springs (Fig. 1.1). In this case m stands for the moment of inertia, and y_n the angle of torsion.

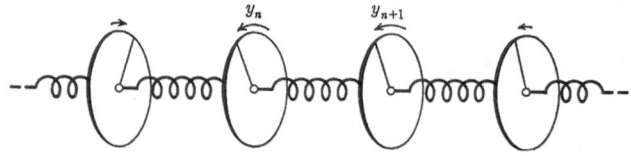

Fig. 1.1.   Models for one-dimensional lattices

If $y_n^{(1)} = y_n^{(1)}(t)$ and $y_n^{(2)} = y_n^{(2)}(t)$ are solutions of (1.1.5), then the linear superposition

$$y_n = y_n^{(1)} + y_n^{(2)}$$ (1.1.6)

is also a solution of the linear equation (1.1.5). Especially when the particles $n = 0$ and $n = N + 1$ are fixed,

$$\left.\begin{aligned} y_n^{(l)}(t) &= C_n \sin\left(\frac{\pi l}{N+1} n\right) \cos\left(\omega_l t + \delta_l\right) \\ \omega_l &= 2\sqrt{\frac{\kappa}{m}} \sin\frac{\pi l}{2(N+1)} \quad (l = 1, 2, \cdots, N) \end{aligned}\right\}$$ (1.1.7)

is the lth normal mode, and the general motion is given by a linear superposition of such modes. The amplitude C_n of each mode is a constant determined by the initial conditions, and no energy transfer occurs between the modes. The linear lattice is therefore nonergodic, and cannot be an object of statistical mechanics unless some modification of the system is made.

We do not enter the ergodic problem here. However, to approach ergodicity we may modify the system by (I) giving nonlinearity to the interaction, (II) introducing impurities, or (III) raising dimensionality. We may have the following comments: (I) In real crystals phonons scatter each other by nonlinearity, which is thought to be responsible for finite thermal conductivity at high temperatures. Usual theories, including that of Peierls, are insufficient from the fundamental point of view because they assume irreversibility from the beginning. (II) Impurities may have some effect similar to absorption and emission of electromagnetic waves by the atom introduced by Planck in the discussion of thermal radiation. (III) two-dimensional and three-dimensional systems belong to future problems. However, as we see later, even in one-dimensional systems nonlinearity and impurities may lead to apparently ergodic behavior.

Fermi did some work on the ergodic problem when he was young, and when electronic computers were developed he came back to this as one of the problems computers might solve. He thought that if one added a nonlinear term to the force between particles in a one-dimensional lattice, energy would flow from mode to mode eventually leading the system to a statistical equilibrium state where the energy is shared equally among linear modes (equipartition of energy). *Fermi* et al. [1.1] wanted to verify this expectation by computer experiments.

They tested potentials, one with a cubic term (α is the nonlinearity constant)

$$\phi(r) = \frac{\kappa}{2} r^2 + \frac{\kappa\alpha}{3} r^3 , \tag{1.1.8a}$$

another with a quartic term (α' is the nonlinearity constant)

$$\phi(r) = \frac{\kappa}{2} r^2 + \frac{\kappa\alpha'}{4} r^4 , \tag{1.1.8b}$$

and the third one with broken linear force ($\kappa \neq \kappa'$, both positive const.)

$$f(r) = \begin{cases} -\kappa r & (|r| < r_0) \\ -(\kappa - \kappa') r_0 - \kappa' r & (r_0 < |r|) \end{cases} . \tag{1.1.8c}$$

For these potentials the results turned out to be qualitatively similar.

They treated lattices with $N = 32$ and $N = 64$ particles, and both ends ($n = 0$ and $n = N + 1$) were fixed. The lattice was initially at rest and given the displacement

$$y_n(0) = B \sin \frac{\pi n}{N + 1} . \tag{1.1.9}$$

That is, they excited the lowest mode. An example of the time change in energy of each mode is shown in Fig. 1.2. As one sees, only a small amount (about 10%), flew from E_1 to E_2, E_3 and so forth, and after a certain time almost all the energy went back to the initial mode. The displacement of each particle also went back to the initial state as is shown in Fig. 1.3. This is the so-called FPU recurrence phenomenon.

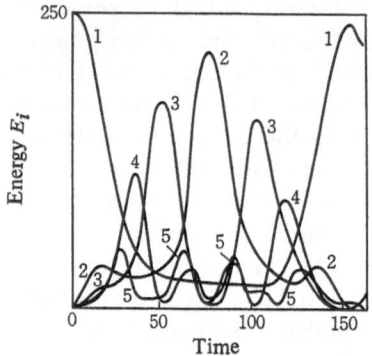

Fig. 1.2. Recurrence phenomena (FermiPasta-Ulam)

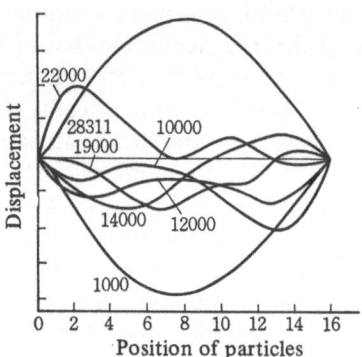

Fig. 1.3. Recurrence of the wave form. Numbers indicate time (unit is different from Fig. 1.2)

Computer experiments sometimes yield unexpected findings, and the FPU recurrence phenomenon is one of them. This was verified by many researchers [1.2–4]. It can be said that if the energy is not large and if the initial wave form is smooth enough, recurrence phenomena will occur. *Zabusky* [1.5] summarized the results by an empirical formula, which says that, when we initially excite the lowest mode, the recurrence time t_R is given by

$$t_R = 0.44 \frac{N^{3/2}}{\sqrt{|\alpha| B}} t_L \qquad (1.1.10)$$

where t_L is the time in which a wave of long wavelength travels forward and back along a lattice of N particles with both ends fixed (or the time necessary for a wave to go round a cyclic lattice of $2N$ particles), that is,

$$t_L = 2N/\sqrt{\kappa/m} . \qquad (1.1.11)$$

We can explain (1.1.10) in terms of the motion of solitons [1.6] (cf. Sect. 2.7).

Saito (cf. Appendix H) numerically found that energy partition among modes takes place abruptly after a time, called the induction period, if one starts from higher modes (for example, $l = 11$). This tendency is clearly shown when the energy exceeds a certain threshold value, which depends on the initial mode. Similar results have also been obtained by Russian workers [1.7]. Saito showed that such results are explained by considering coupling between modes. The recurrence phenomenon FPU obtained seems to be due to the fact that the initial conditions were sufficiently smooth and the total energy was not large compared with nonlinearity. We must note that though energy is equally shared if the system is ergodic, energy sharing does not guarantee ergodicity. If particle energy is distributed according to Maxwell's law, it has no relation to ergodicity. As a matter of fact, it is shown for a linear lattice that almost all the initial conditions lead immediately to Maxwell's distribution of particle energy. However, the correlation of motion of particles remains forever.

FPU took the systems with $N = 32$ and $N = 64$. In these cases the eigenfrequencies given by (1.1.7) are mutually prime, that is, for any set of nonzero n_i's we have $\sum n_i \omega_i \neq 0$, which implies nonresonance. *Ford* and co-workers [1.2] showed by using perturbation and by numerical calculation that, though resonance generally enhances energy sharing, it has no intimate connection to recurrence phenomena, and that nonlinear lattices have rather stable motion, which he called the nonlinear normal modes. This remarkable property led to the finding of an integrable 1-dimensional lattice with exponential interaction. Before introducing this lattice in the next chapter, some related subjects are to be supplemented in the following sections.

1.2 Hénon-Heiles Calculation

In addition to what was mentioned at the end of the last section, Lunsford and Ford [1.8] found another remarkable fact. As the simplest system, we frequently refer to three-particle periodic systems. The Hamiltonian for a linear periodic lattice of three particles can be written in a dimensionless form as

$$\mathcal{H}_0 = \frac{1}{2}(P_1^2 + P_2^2 + P_3^2) + \frac{1}{2}[(Q_1 - Q_2)^2 + (Q_2 - Q_3)^2 + (Q_3 - Q_1)^2]$$

$$(1.2.1)$$

where P stands for the momentum and Q the displacement. Since the total momentum is conserved the number of freedom can be reduced to 2. For this purpose, we introduce normal coordinates ζ_1, ζ_2 and ζ_3 with momenta η_1, η_2 and η_3, respectively, in such a way that ζ_3 is the cyclic coordinate. We thus have a rotation in the phase space

$$\left.\begin{array}{l} Q_i = \sum_{j=1}^{3} A_{ij}\zeta_j \\[2mm] P_i = \sum_{j=1}^{3} A_{ij}\eta_j \end{array}\right\},$$

$$(1.2.2a)$$

where

$$A = \begin{pmatrix} 6^{-1/2} & 2^{-1/2} & 3^{-1/2} \\ -(2/3)^{1/2} & 0 & 3^{-1/2} \\ 6^{-1/2} & -2^{-1/2} & 3^{-1/2} \end{pmatrix}. \tag{1.2.2b}$$

Then the Hamiltonian is transformed to

$$\mathscr{H}_0 = \frac{1}{2}(\eta_1^2 + \eta_2^2 + \eta_3^2) + \frac{3}{2}(\zeta_1^2 + \zeta_2^2). \tag{1.2.3}$$

When we add cubic nonlinear interaction the Hamiltonian becomes

$$\mathscr{H} = \mathscr{H}_0 + \frac{1}{3}\alpha[(Q_1 - Q_2)^3 + (Q_2 - Q_3)^3 + (Q_3 - Q_1)^3] \tag{1.2.4}$$

and the above transformation yields

$$\mathscr{H} = \frac{1}{2}(\eta_1^2 + \eta_2^2 + \eta_3^2 + 3\zeta_1^2 + 3\zeta_2^2) + \frac{3\alpha}{2^{1/2}}\left(\zeta_2\zeta_1^2 - \frac{1}{3}\zeta_2^3\right). \tag{1.2.5}$$

By the scale transformation

$$\zeta_1 = \frac{2^{1/2}}{\alpha}q_1, \quad \zeta_2 = \frac{2^{1/2}}{\alpha}q_2$$
$$t \to t/3^{1/2}, \quad \mathscr{H} \to \frac{6}{\alpha^2}\mathscr{H} \tag{1.2.6}$$

we have the Hamiltonian ($p_1 = \dot{q}_1, p_2 = \dot{q}_2$)

$$\mathscr{H} = \frac{1}{2}(p_1^2 + p_2^2) + \frac{1}{2}(q_1^2 + q_2^2) + q_1^2 q_2 - \frac{1}{3}q_2^3 \tag{1.2.7}$$

which expresses the two-dimensional motion of a mass point.

Equation (1.2.7) is the Hamiltonian of a model used by *Hénon* and *Heiles* (HH) [1.9] to study the motion of a star in a galaxy with cylindrical symmetry. They numerically integrated the equations of motion derived from (1.2.7), that is,

$$\dot{q}_1 = \frac{\partial \mathscr{H}}{\partial p_1} = p_1$$

$$\dot{q}_2 = \frac{\partial \mathscr{H}}{\partial p_2} = p_2 \tag{1.2.8}$$

$$\dot{p}_1 = -\frac{\partial \mathcal{H}}{\partial q_1} = -q_1 - 2q_1 q_2$$

$$\dot{p}_2 = -\frac{\partial \mathcal{H}}{\partial q_2} = -q_2 - q_1^2 + q_2^2 .$$

The motion can be expressed by a trajectory in the four-dimensional phase space (q_1, q_2, p_1, p_2). The figure made by the point where trajectories cut a two-dimensional plane in the phase space is called Poincaré mapping or Poincaré surface of section. When the energy is small mapping makes smooth curves as is shown in Fig. 1.4a. In this case, each trajectory is on a surface which is different from the surface of constant energy indicating the existence of an integral (the so-called third integral), besides the energy integral. In such a case, the system is said to be integrable. The system is then nonergodic. However, when the energy exceeds a certain critical value, there appear some regions where the mapping of a trajectory looks quite erratic (apparently stochastic). In Fig. 1.4b many points belong to one trajectory. If we plot the area of such a stable region against

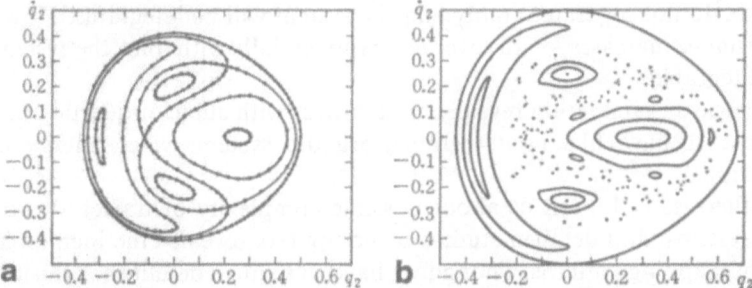

Fig. 1.4a,b. Poincaré mapping of Hénon-Heiles (a) energy $E = 0.08333$; (b) energy $E = 0.12500$

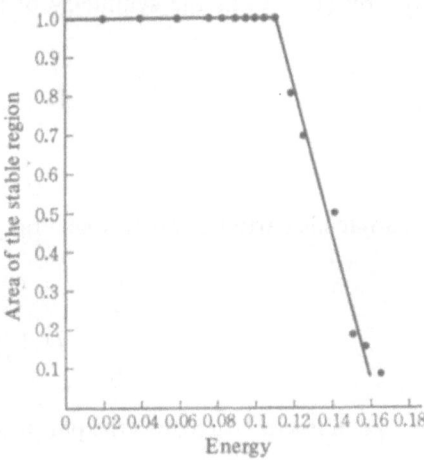

Fig. 1.5. Relative area of the stable region of Hénon-Heiles system

energy, we have Fig. 1.5. $E_c \simeq 0.11$ is the threshold energy. *Contropoulos* [1.10], *Hénon*, and *Saito* found similar behavior of other systems with a low degree of freedom. Such erratic behavior is not yet fully understood [1.11].

Since a three-particle periodic lattice with a cubic nonlinear term is equivalent to an HH-system, we may expect that when the energy is lower than a threshold value, the motion in the lattice is accessible to analysis, the system being integrable. However, when the energy exceeds this threshold, erratic regions appear, and when the energy becomes sufficiently large all the regions in the phase space will become ergodic.

Mapping is difficult for a lattice of many more particles. Instead, Ford studied the temporal change of the distance between two points initially very close in the phase space. If two points are specified by $(Q_1^{(1)}, Q_2^{(1)}, ..., P_1^{(1)}, P_2^{(1)}, ...)$ and $(Q_1^{(2)}, Q_2^{(2)}, ..., P_1^{(2)}, P_2^{(2)}, ...)$, then the distance is

$$\sqrt{\sum_i [(Q_i^{(2)} - Q_i^{(1)})^2 + (P_i^{(2)} - P_i^{(1)})^2]} . \tag{1.2.9}$$

Though this value oscillates, if, on the average, it increases linearly with time, the two points do not separate rapidly and the system will be integrable. However, if the distance increases, on the average, exponentially with time, the system will be nonintegrable.

Such numerical examination revealed that lattices with cubic or quartic nonlinear potential behave like nonintegrable, ergodic systems when energy is sufficiently large.

Ergodic theories [1.12] tell us about possible complexity of trajectories of mechanical systems, and detailed studies by computers revealed the incredible complexity of mapping patterns. Though we have to retain a definite conclusion on this problem, contributions of computer experiments to the ergodic problem will have increasing importance in the future.

Problem 1.1. Show that the potential given by (1.2.7) has the symmetry of a regular triangle.

1.3 Discovery of Solitons

For small change r of the distance between adjacent particles, we may generally expand the interaction potential as

$$\phi(r) \simeq \frac{\kappa}{2} r^2 + \frac{\kappa}{3} \alpha r^3 + \cdots . \tag{1.3.1}$$

If we put $\alpha = 0$ and neglect higher terms we get a linear lattice. α is the parameter of nonlinearity. Further, for smooth waves with slow change of r, we may approximate a lattice by a continuum. If α is very small, equations of motion can be approximated by (1.1.5), and for very smooth waves, writing $y_n(t) = y(x, t)$ in (1.1.5), the right-hand side can be approximated by $\kappa h^2(\partial^2 y/\partial x^2)$, where h is the average distance between adjacent particles, and $x = nh$ is the position of the particle. In this continuum approximation (1.1.5) reduces to the wave equation $m(\partial^2 y/\partial t^2) = \kappa h^2(\partial^2 y/\partial x^2)$, and waves propagate to the right or to the left with the speed

$$c_0 = h\sqrt{\kappa/m}\,. \tag{1.3.2}$$

For waves going to the right we introduce a moving coordinate

$$\xi = nh - c_0 t \tag{1.3.3}$$

and retain the lowest nonlinear term. Then we can describe waves by the nonlinear partial differential equation (cf. Sect. 2.7)

$$\frac{\partial u}{\partial \tau} + u\frac{\partial u}{\partial \xi} + \delta^2 \frac{\partial^3 u}{\partial \xi^3} = 0\,, \tag{1.3.4}$$

where we assume cubic nonlinearity ($\alpha \neq 0$), u is a quantity proportional to r, and τ is a time variable proportional to t. Equation (1.3.4) was proposed by *Korteweg* and *de Vries* [1.13] in 1895 to describe shallow water waves, and is called the Korteweg-de Vries (KdV) equation.

Around 1965 it was pointed out that certain plasma waves are described by the KdV equation, and this equation was numerically studied since an analytic general solution was not found. Thus, *Zabusky* and *Kruskal* [1.14] discovered that waves subject to the KdV equation were composed of particular pulses. Each pulse is a particular solution of the KdV equation and has the form

$$\begin{aligned} u_s &= u_0 + A\,\mathrm{sech}^2[(\xi - ct)/\Delta] \\ \Delta &= \delta\sqrt{12/A}, \qquad c = u_0 + A/3 \end{aligned} \tag{1.3.5}$$

It was found that when two such pulses collided they went through one another (or absorbed and emitted) to recover their initial shapes. Since the pulse had such a property similar to a stable particle, they named it a soliton, which means a solitary wave particle.

As will be described later, we also have solitons in nonlinear lattices. Around 1965 *Visscher* et al. [1.15] numerically studied wave propagation in a nonlinear lattice, and he already saw solitons in his results though we could only recognize them later. He used Lennard-Jones type potentials and studied energy transfer or thermal conductivity. For one- and two-dimensional lattices with many impurities of different masses, he found that energy transfer was generally enhanced by the introduction of nonlinear interaction terms. Such results can be understood by the fact that energy is accumulated in the form of solitons, which propagate without being hindered very much by impurities.

1.4 Dual Systems

The lattice with exponential interaction, the main subject of this book, was found after looking for a system with rigorous periodic solution which had been expected from numerical works by Ford. Then the concept of dual systems was quite useful. Systems A and B are said to be dual to one another, if B is obtained from A by repalcing particles by springs and springs by particles following certain rules. For a harmonic lattice we can replace heavier (lighter) particles by weaker (stronger) springs in such a way that the normal mode frequencies are the same in both systems [1.16]. Then they are dual to one another (Fig. 1.6).

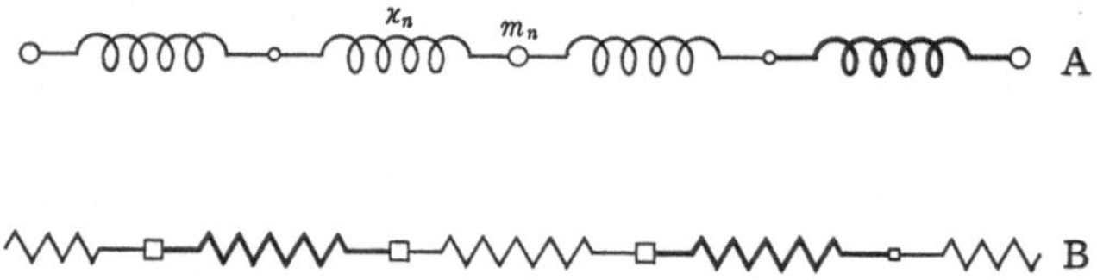

Fig. 1.6. Relation between dual systems

We can generalize the idea of dual systems by the following consideration. The Hamiltonian which gives rise to the equation of motion (1.1.1) is [1.17]

$$\mathscr{H} = \frac{1}{2m} \sum_n p_n^2 + \sum_n \phi(r_n) \,, \tag{1.4.1}$$

where the momentum p_n is related to the kinetic energy

$$K = \frac{1}{2} \sum_n m\dot{y}_n^2 \tag{1.4.2}$$

by differentiation of K with respect to the velocity $\dot{y}_n$:

$$p_n = \frac{\partial K}{\partial \dot{y}_n} = m\dot{y}_n \,. \tag{1.4.3}$$

Now, we use the mutual displacement r_n as the generalized coordinate. For brevity's sake we assume that the left end particle $n = 0$ is fixed. Then we have

$$y_0 = 0, \quad y_1 = r_0, \quad y_2 = r_0 + r_1, \cdots$$
$$\dot{y}_0 = 0, \quad \dot{y}_1 = \dot{r}_0, \quad \dot{y}_2 = \dot{r}_0 + \dot{r}_1, \cdots \tag{1.4.4}$$

and for a lattice with N movable particles

$$K = \frac{1}{2} \sum_{n=0}^{N-1} m(\dot{r}_0 + \dot{r}_1 + \cdots + \dot{r}_n)^2 . \tag{1.4.5}$$

The momentum s_n conjugate to r_n is defined by

$$s_n = \frac{\partial K}{\partial \dot{r}_n} = m \left[(\dot{r}_0 + \dot{r}_1 + \cdots + \dot{r}_n) + (\dot{r}_0 + \dot{r}_1 + \cdots + \dot{r}_{n+1}) + \cdots \right.$$
$$\left. + (\dot{r}_0 + \dot{r}_1 + \dot{r}_2 + \cdots + \dot{r}_{N-1}) \right] . \tag{1.4.6}$$

Therefore we have

$$s_{n-1} - s_n = m\dot{y}_n \tag{1.4.7a}$$

$$s_N = 0 \tag{1.4.7b}$$

and the Hamiltonian becomes

$$\mathscr{H} = \frac{1}{2m} \sum_{n=0}^{N-1} (s_{n+1} - s_n)^2 + \sum_{n=0}^{N-1} \phi(r_n) . \tag{1.4.8}$$

The canonical equations of motion are

$$\dot{r}_n = \frac{\partial \mathscr{H}}{\partial s_n} = -\frac{s_{n+1} - 2s_n + s_{n-1}}{m} \tag{1.4.9}$$

$$\dot{s}_n = -\frac{\partial \mathscr{H}}{\partial r_n} = -\phi'(r_n) . \tag{1.4.10}$$

If we eliminate s_n from these equations we obtain

$$m\ddot{r}_n = \phi'(r_{n+1}) - 2\phi'(r_n) + \phi'(r_{n-1}) \tag{1.4.11}$$

which is, however, the difference of (1.1.1) and another equation in which n is replaced by $n + 1$, and therefore is not a new equation.

If (1.4.10) admits the inverse, we may write

$$r_n = -\frac{1}{m} \chi(\dot{s}_n) . \tag{1.4.12}$$

Then we can eliminate r_n from (1.4.9) to obtain

$$\frac{d}{dt} \chi(\dot{s}_n) = s_{n+1} - 2s_n + s_{n-1} \ .$$
(1.4.13)

This is an equation dual to (1.4.11). If we think of s_n as the "displacement", then the right-hand side of (1.4.13) can be interpreted as the force of linear springs, and in the left-hand side $\chi(\dot{s}_n)$ can be interpreted as the momentum associated to the "speed" $\dot{s}_n$. Then (1.4.13) turns out to be mechanical equations of motion.

The force f_s of the spring is related to $\dot{s}_n$ by

$$f_s = -\phi'(r_n) = \dot{s}_n$$
(1.4.14)

and the equation of motion (1.4.11) is rewritten as

$$\frac{d^2}{dt^2} \chi(f_n) = f_{n+1} - 2f_n + f_{n-1} \ .$$
(1.4.15)

Further, we introduce the integral of s_n by

$$S_n = \int^t s_n dt \ .$$
(1.4.16)

Choosing the integration constant appropriately, we have from (1.4.7a, 9)

$$y_n = \frac{1}{m} (S_{n-1} - S_n)$$
(1.4.17)

$$r_n = \frac{-1}{m} (S_{n+1} - 2S_n + S_{n-1}),$$
(1.4.18)

and the equations of motion take the form

$$\chi(\ddot{S}_n) = S_{n+1} - 2S_n + S_{n-1} \ .$$
(1.4.19)

Problem 1.2. Assume that the masses and the spring constants of two harmonic lattices, (m_n, κ_n) and (m_n', κ_n'), respectively, satisfy

$$m_n \kappa_n' = m_n' \kappa_n = \text{const. (independent of } n).$$

Then, show that all the normal mode frequencies of these systems coincide. Further, show that a free (fixed) end of a system corresponds to a fixed (free) end of the dual system.

Hint: Write down the Hamiltonians.

Problem 1.3. Express the equations of motion in terms of S_n, when the interaction potential $\phi(r)$ is hyperbolic, $\phi(r) = \alpha\sqrt{\beta + r^2}$ (α, β are constants).

Remark: In this case the momentum takes relativistic form.

Problem 1.4. Confirm that if a constant external pressure is applied to the lattice,

$$\dot{s}_n = -\phi'(r_n) - f$$

and (1.4.13) becomes

$$\frac{d}{dt}\chi(\dot{s}_n + f) = s_{n+1} - 2s_n + s_{n-1}.$$

Hint: The potential energy $f \cdot (r_0 + r_1 + r_2 + \ldots r_N)$ must be added to (1.4.8).

2. The Lattice with Exponential Interaction

In the preceding chapter, it has been shown that nonlinear lattices studied by Fermi et al. have periodic behavior at least when the energy is not too high, and that stable pulses (solitons) propagate in nonlinear continuous systems. These facts indicate that there will be some nonlinear lattice which admits rigorous periodic waves, and that certain pulses (lattice solitons) will be stable there. In this chapter, we seek such a lattice, and will show that a lattice with exponential interaction admits rigorous periodic solutions and soliton solutions. Two-soliton solutions are also given. A system of hard spheres as the limit of sharp repulsive forces, and the continuum limit for smooth waves are examined. Further, an application of the theory of the lattice to a nonlinear electric circuit, and some extensions are discussed. Finally, it is pointed out that the lattice with exponential interaction has many conserved quantities besides momentum and energy, and this leads to the next chapter.

2.1 Finding of an Integrable Lattice

To elucidate characteristic features of waves in nonlinear lattices, we expected a nonlinear lattice which admits particular solutions with wide applicability. This means, to look for potential function $\phi(r)$ which admits integration of the equations of motion (1.1.1). Equation (1.1.1), or equivalently (1.4.11), is quite familiar, but it seemed hard to find such a potential $\phi(r)$ from this. On the contrary, (1.4.13) can be considered as a recurrence formula by which s_{n+1} is derived if we have s_{n-1} and s_n, including a derivative of a function of s_n, which is related to the inverse function of the potential $\phi(r)$. It is also required that the potential must have some physical meaning, so that it really provides us with a mechanical system with wide applicability. Under these conditions, many functions s_n and $\phi(r)$ were tried to find out the combination by which (1.4.13) was satisfied [2.1].

In the case of a harmonic lattice, typical periodic waves are sinusoidal or trigonometric. Therefore it was quite natural to think of elliptic functions as possible candidates, because they are in a sense extensions of trigonometric functions. However, waves proportional to Jacobian sn or cn functions could not give the right answer.

Meanwhile we noticed an addition formula for sn^2

$$\text{sn}^2(u+v) - \text{sn}^2(u-v) = 2\frac{d}{dv}\left(\frac{\text{sn } u \text{ cn } u \text{ dn } u \text{ sn}^2v}{1 - k^2 \text{ sn}^2u \text{ sn}^2v}\right) \tag{2.1.1}$$

which led to the lattice being searched for.

Now, using

$$\text{dn}^2u = 1 - k^2 \text{ sn}^2u \tag{2.1.2}$$

we define a function $\varepsilon(u)$ by

$$\varepsilon(u) = \int_0^u (\text{dn}^2u) \, du \tag{2.1.3}$$

to have

$$\begin{aligned} \varepsilon'(u) &= \text{dn}^2u \\ \varepsilon''(u) &= - 2k^2 \text{ sn } u \text{ cn } u \text{ dn } u \end{aligned} \tag{2.1.4}$$

and

$$\varepsilon(u+v) + \varepsilon(u-v) - 2\varepsilon(u) = \frac{\varepsilon''(u)}{1/\text{sn}^2v - 1 + \varepsilon'(u)}. \tag{2.1.5}$$

Though $\varepsilon(u)$ is not a periodic function, a Jacobian zn function (ζ function) defined by

$$Z(u) = \varepsilon(u) - \frac{E}{K} u \tag{2.1.6}$$

is a periodic function with the period $2K$, where K and E are, respectively, the complete elliptic integrals of the first and second kind.

Rewriting (2.1.5) we have

$$Z(u+v) + Z(u-v) - 2Z(u) = \frac{d}{du}\ln\left[1 + \frac{1}{1/\text{sn}^2v - 1 + E/K}Z'(u)\right] \tag{2.1.7}$$

which is to be compared with (1.4.13). Thus we see that (1.4.13) is satisfied when we put

$$\begin{aligned} u &= 2\left(vt \pm \frac{n}{\lambda}\right)K \\ v &= 2K/\lambda \end{aligned} \tag{2.1.8}$$

where λ (the wavelength) and ν (the frequency) are constants, and identify the function s_n and χ with

$$s_n(t) = \frac{2K\nu}{b/m} Z(u) \tag{2.1.9}$$

and

$$\chi(\dot{s}) = \frac{m}{b} \ln\left[1 + \frac{\frac{b/m}{(2K\nu)^2}}{1/\mathrm{sn}^2 v - 1 + E/K} \dot{s}\right] - m\sigma, \tag{2.1.10}$$

where b and σ are constants. $\chi(\dot{s})$ is the inverse function of $\dot{s} = \phi(r)$, the potential, and must not contain ν and v, which means that the factor of $\dot{s}$ in (2.1.10) is a constant independent of ν and v. Therefore the relation

$$(2K\nu)^2 = \frac{ab}{m}\left[\frac{1}{\mathrm{sn}^2(2K/\gamma)} - 1 + \frac{E}{K}\right]^{-1} \tag{2.1.11}$$

must hold, where a is a constant, and in order that the right-hand side is positive, we must assume that $ab > 0$.

By (1.4.12) and (2.1.10), we have

$$r = -\frac{1}{b} \ln\left(1 + \frac{\dot{s}}{a}\right) + \sigma \tag{2.1.12}$$

with $r = r_n$. Taking the inverse, by (1.4.10) we have

$$\dot{s} = a[e^{-b(r-\sigma)} - 1] = -\phi'(r). \tag{2.1.13}$$

Therefore, as the potential we obtain a function with three parameters a, b, and σ, which can be written as

$$\phi(r) = \frac{a}{b} e^{-b(r-\sigma)} + ar + \text{const.} \tag{2.1.14}$$

or

$$\phi(r) = Ae^{-br} + ar. \tag{2.1.15}$$

Problem 2.1. Calculate $\chi(\dot{s})$ of (1.4.12) for the potential given by (2.1.14).

2.2 The Lattice with Exponential Interaction

If we take the position of the minimum of $\phi(r)$ as the origin $r = 0$, the potential (2.1.14) takes the form

$$\phi(r) = \frac{a}{b} e^{-bt} + ar \qquad (ab > 0). \tag{2.2.1}$$

In the following, we use this simple expression for the interaction potential. The lattice with the exponential interaction is often referred to as the Toda lattice.

The potential (2.2.1) for $a, b > 0$ has strong repulsion and weak attraction as shown by Fig. 2.1a. Near the minimum it gives a harmonic force, and with increasing distance nonlinearity of the force comes in. Thus it has the nature of physical atomic forces. For $a, b < 0$, the potential has weak repulsion and strong attraction as shown by Fig. 2.1b. In what follows we assume the case $a, b > 0$, but we may similarly discuss the case $a, b < 0$.

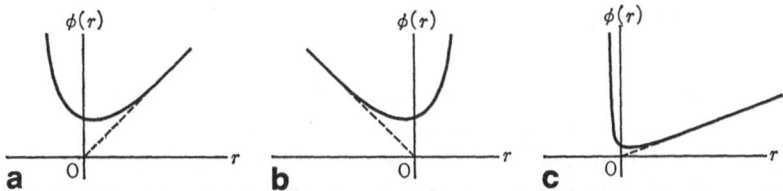

Fig. 2.1a-c. Exponential potential   (a) $a,b > 0$, (b) $a,b < 0$, (c) when b is quite large

If we expand (2.2.1) assuming small r, we have

$$\phi(r) = \text{const.} + \frac{ab}{2} r^2 - \frac{ab^2}{6} r^3 + \cdots . \tag{2.2.2}$$

Thus, for sufficiently small motion, the lattice looks like a linear lattice with the spring constant

$$\kappa = ab . \tag{2.2.3}$$

For a somewhat larger motion the nonlinear parameter of (1.1.8) is given by

$$\alpha = -\frac{b}{2} . \tag{2.2.4}$$

In the limit of very large b, we have a system of hard spheres (rods). The lattice with exponential interaction is a model system which has a harmonic lattice and a hard-sphere system as both limits, and it has the merit that if we obtain a particular solution it applies also to these limits.

When an external pressure f is applied to the lattice, since it is equivalent to an additional potential fr, we have only to replace the interaction by

$$\phi(r) = \frac{a}{b} e^{-br} + (a + f)r , \tag{2.2.5}$$

where $f > 0$ is a pressure and $f < 0$ a tension. In regard to (2.1.15), we may think, for a one-dimensional lattice, that the term ar is due to a constant external force and the interaction is only the repulsive potential $A \exp(-br)$. Further, in (2.2.5), $a + f$ can be considered as an attraction force; and when $f = \text{const.}$, introducing constants a', σ' such that

$$\left.\begin{array}{l} a' = a + f \\ a'e^{-b\sigma'} = a \end{array}\right\},$$

(2.2.6)

we may write (2.2.5) as

$$\phi(r) = \frac{a'}{b} e^{-b(r+\sigma')} + a'r .$$

(2.2.7)

This means that replacing a by a' and shifting the origin of r by $-\sigma'$, the external force can be taken into the exponential type of interaction.

In any case, the second term of (2.2.1) gives constant forces on each particle from the left and the right neighbors, which therefore cancel out and do not appear explicitly in the equations of motion. Indeed, assuming the potential (2.2.1), we have from (1.1.1) and (1.4.11) the equations of motion

$$m\frac{d^2y_n}{dt^2} = a[e^{-b(y_n - y_{n-1})} - e^{-b(y_{n+1} - y_n)}]$$

(2.2.8)

$$m\frac{d^2r_n}{dt^2} = a(2e^{-br_n} - e^{-br_{n-1}} - e^{-br_{n+1}}) .$$

(2.2.9)

For the equivalent dual expression, (1.4.13) yields

$$\frac{d}{dt} \ln(a + \dot{s}_n) = \frac{b}{m}(s_{n-1} - 2s_n + s_{n+1})$$

(2.2.10)

or

$$\frac{\ddot{s}_n}{a + \dot{s}_n} = \frac{b}{m}(s_{n-1} - 2s_n + s_{n+1}) .$$

(2.2.11)

Differentiating the last equation, we have

$$\frac{d^2}{dt^2} \ln\left(1 + \frac{f_n}{a}\right) = \frac{b}{m}(f_{n-1} - 2f_n + f_{n+1}) ,$$

(2.2.12)

and integrating we obtain

$$\ln\left(1 + \frac{\ddot{S}_n}{a}\right) = \frac{b}{m}(S_{n-1} - 2S_n + S_{n+1})$$

(2.2.13)

after choosing the integration constants appropriately [2.1a, 2]. These are equations of motion for the lattice with exponential interaction. The force of the spring is, by (1.4.14), given as

$$f_n = a(e^{-br_n} - 1) = \dot{s}_n \,. \tag{2.2.14}$$

In the following sections, we investigate some particular solutions.

2.3 Periodic Solutions

We have a periodic solution as given by (2.1.9). Differentiating this with respect to time and referring to (2.2.14) and (2.1.6), we have a periodic solution

$$e^{-br_n} - 1 = \frac{(2K\nu)^2}{ab/m} \left\{ \mathrm{dn}^2 \left[2 \left(\frac{n}{\lambda} \pm \nu t \right) K \right] - \frac{E}{K} \right\}, \tag{2.3.1}$$

where the wavelength λ is arbitrary, and the frequency ν is related to λ by (2.1.11); or

$$2K\nu = \sqrt{\frac{ab}{m}} \Big/ \sqrt{\frac{1}{\mathrm{sn}^2(2K/\lambda)} - 1 + \frac{E}{K}} \tag{2.3.2}$$

which is the dispersion relation for this wave. In the above equations K is the complete elliptic integral of the first kind and E the second kind, which are defined as

$$K = K(k) = \int_0^{\pi/2} \frac{d\theta}{\sqrt{1 - k^2 \sin^2\theta}} = \int_0^1 \frac{dx}{\sqrt{(1 - x^2)(1 - k^2 x^2)}} \tag{2.3.3a}$$

$$E = E(k) = \int_0^{\pi/2} \sqrt{1 - k^2 \sin^2\theta} \; d\theta = \int_0^1 \sqrt{\frac{1 - k^2 x^2}{1 - x^2}} \, dx \,. \tag{2.3.3b}$$

When the modulus k is small, the amplitude of the periodic wave is small and the wave is very similar to a sinusoidal wave. When k is large being nearly unity, the crest of the wave becomes steep and the trough becomes very flat (which is similar to water waves).

The Jacobian function dn can be written in terms of cn. As mentioned in Sect. 1.3, in the continuum approximation, lattice waves can be approximated by those of the KdV equation, and the periodic solution to this equation obtained by Korteweg and de Vries was expressed in terms of cn function and they used the term cnoidal wave (as a sine wave is called a sinusoidal wave). It can be shown that, for sufficiently small amplitude, the periodic wave (2.3.1) coincides with the cnoidal wave of the KdV equation. Thus, the periodic wave (2.3.1) is called the lattice cnoidal wave.

When the modulus k is close to zero, we have $E/K \simeq 1 - k^2/2$. Therefore (2.3.1, 2) give the approximation

$$r_n \simeq -\frac{\omega^2 k^2}{8ab^2} \cos\left(\omega t \pm \frac{2\pi n}{\lambda}\right) \qquad (2.3.4a)$$

$$\omega = 2\pi\nu \simeq \sqrt{\frac{ab}{m}}\Big/\lambda \qquad (2.3.4b)$$

which is a sinusoidal wave.

Writing the right-hand side of (2.3.1) by a Fourier series we have

$$dn^2(2xK) - \frac{E}{K} = \frac{\pi^2}{K^2} \sum_{l=1}^{\infty} \frac{l \cos 2\pi l x}{\sinh(\pi l K'/K)}, \qquad (2.3.5)$$

where we have introduced $x = n/\lambda \pm \nu t$, and in terms of the complementary modulus k' we have defined K' by

$$K' = K(k'), \qquad k' = \sqrt{1 - k^2}. \qquad (2.3.6)$$

The above periodic function can also be written as

$$dn^2(2xK) - \frac{E}{K} = \left(\frac{\pi}{2K'}\right)^2 \sum_{l=-\infty}^{\infty} \operatorname{sech}^2\left[\frac{\pi K}{K'}(x - l)\right] - \frac{\pi}{2KK'} \qquad (2.3.7)$$

and the cnoidal wave can be written as

$$e^{-br_n} - 1 = \frac{m}{ab}\left\{\sum_{l=-\infty}^{\infty} \beta^2 \operatorname{sech}^2\left[\alpha(n - \lambda l) - \beta t\right] - 2\beta\nu\right\} \qquad (2.3.8a)$$

with

$$\alpha = \frac{\pi K}{\lambda K'}, \qquad \beta = \frac{\pi K \nu}{K'}. \qquad (2.3.8b)$$

Equation (2.3.8a) says that a cnoidal wave can be seen as a succession of pulses (sech^2), the distance being the wavelength λ.

An example of the waveform and the dispersion relation for a cnoidal wave is shown in Figs. 2.2 and 2.3. The dispersion relation looks similar to that for a harmonic lattice.

The Z function introduced in (2.1.6) is related to the ϑ_0 function by

$$2KZ(2xK) = \frac{d}{dx}\ln\vartheta_0(x) \qquad (2.3.9)$$

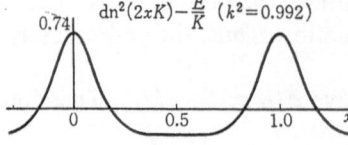

Fig. 2.2. Cnoidal wave form

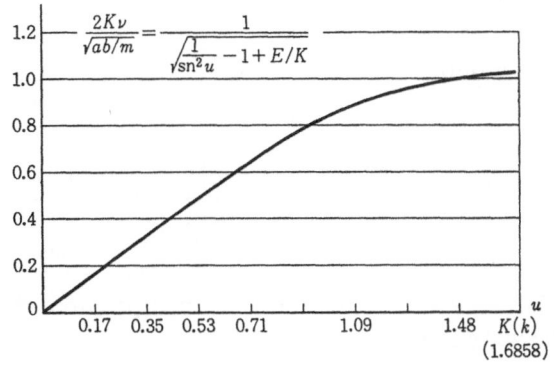

Fig. 2.3. Dispersion relation of a cnoidal wave ($k = 0.5$)

and a cnoidal wave can be expressed in terms of ϑ_0. Writing the result for s_n, from (2.1.9), we have

$$s_n = \frac{1}{b/m} \frac{d}{dt} \ln \vartheta_0 \left(\nu t \pm \frac{n}{\lambda} \right). \tag{2.3.10}$$

Therefore, (1.4.7a) gives

$$m\dot{y}_n = \frac{m}{b} \frac{d}{dt} \ln \frac{\vartheta_0[\nu t \pm (n-1)/\lambda]}{\vartheta_0(\nu t \pm n/\lambda)} \tag{2.3.11}$$

for the momentum of a particle in a cnoidal wave.

Problem 2.2. Show that when we increase the wave number $1/\lambda$ by any integer, we obtain the same wave for a lattice cnoidal wave.

Hint: The dn function has the period $2K$.

Problem 2.3. In regard to the dispersion relation $\omega = \omega(1/\lambda)$ for a cnoidal wave, show that

$$\omega\left(\frac{1}{\lambda}\right) = \omega\left(1 - \frac{1}{\lambda}\right) = \omega\left(1 + \frac{1}{\lambda}\right) = \cdots.$$

Therefore we have no need to consider waves with wavelengths smaller than twice the mean distance between the particles ($\lambda < 2$).

Problem 2.4. Using the relations

$$\mathrm{dn}\,(\beta t \pm nK) = \mathrm{dn}\left[\beta t + \begin{pmatrix} 0 \\ K \end{pmatrix}\right] \quad \begin{pmatrix} n = \text{even} \\ n = \text{odd} \end{pmatrix}$$

$$\mathrm{dn}\,(\beta t + K) = k'/\mathrm{dn}(\beta t) \qquad (k' = \sqrt{1 - k^2})$$

derive the expression for a cnoidal wave with the minimum wave-length $\lambda = 2$, and show that we can write

$$r_{2n} = \Delta + 2x, \qquad r_{2n+1} = \Delta - 2x,$$

and that x satisfies the equation of motion

$$m\ddot{x} = - 2 a' \sinh 2bx$$

with

$$a' = a\,e^{-b\Delta}\,.$$

Problem 2.5. Show that a cnoidal wave can be written as a superposition of two cnoidal waves with twice the wavelength.

Hint: $\vartheta_0(2x, q^2) = \varphi(q)\,\vartheta_0(x,q)\,\vartheta_0(x \pm 1/2, q)\,.$

Problem 2.6. Show that a cnoidal wave can be written as a superposition of cnoidal waves with a wavelength which is any integral multiple of the original wavelength.

Hint: $\vartheta_0(lx, q^l) = \dfrac{\phi(q^l)}{[\phi(q)]^l}\displaystyle\prod_{s=0}^{l-1} \vartheta_0(x + s/l, q)\,.$

Problem 2.7. Show that a cnoidal wave can be written as a superposition of solitons with a mutual distance equal to the wavelength (but the velocity is modified by the interaction between solitons).

2.4 Solitary Waves [2.1a,2]

Among the various expressions of the equations of motion from (2.2.8–13), probably the most tangible ones are (2.2.11, 13). Assuming a wave which propagates without changing its form, we see that these equations of motion are satisfied by

$$s_n = \pm \frac{\beta m}{b} \tanh (n\kappa \pm \beta t) + \text{const.} \tag{2.4.1}$$

or

$$S_n = \frac{m}{b} \ln \cosh (\kappa n \pm \beta t)\,, \tag{2.4.2}$$

with

$$\beta = \sqrt{\frac{ab}{m}}\, \sinh \kappa\,. \tag{2.4.3}$$

Since linear terms of S_n are arbitrary, we may put

$$S_n = \frac{m}{b} \ln (1 + e^{\pm 2(\kappa n \pm \beta t)}).$$ (2.4.4)

These lead to a solitary wave solution

$$e^{-br_n} - 1 = \frac{m}{ab} \beta^2 \operatorname{sech}^2(\kappa n \pm \beta t),$$ (2.4.5)

which is a pulselike wave with the speed

$$c = \frac{\beta}{\kappa} = \sqrt{\frac{ab}{m}} \frac{\sinh \kappa}{\kappa}$$ (2.4.6)

where the lattice spacing is chosen as the unit of length. The larger the height is, the smaller the width ($\sim 1/\kappa$), and the larger the speed of the wave. The solitary wave always propagates faster than waves of long wavelength with the speed $c_0 = \sqrt{ab/m}$.

Numerical calculations show that the solitary wave given by (2.4.5) is stable. That is, even when there are other waves, ripples for example, the solitary wave will never collapse. As is shown in the next section, when two solitary waves collide, they go through one another and recover their initial shapes. Since they behave like stable particles such solitary waves as given by (2.4.5) are called (lattice) solitons.

As the right-hand side of (2.4.5) is positive we see that $r_n \leq 0$ we (assume the case $b > 0$). In other words the lattice is compressed around a soliton. Thus, a soliton in the lattice with an exponential repulsion force is a compressed wave. Figure 2.4 indicates s_n and $\exp(-br_n) - 1$ for a soliton, and a cnoidal wave which may be regarded as a series of solitons, cf. (2.3.8a).

Due to compression, a soliton has excess mass density. Inserting (2.4.5) into (1.4.17), we see that the displacement y_n of a particle in the soliton is given by

$$y_n = \frac{1}{b} \ln \frac{1 + e^{2(\kappa n - \kappa \pm \beta t)}}{1 + e^{2(\kappa n \pm \beta t)}} + \text{const}.$$ (2.4.7)

Therefore the total compression by a soliton is

$$y_{-\infty} - y_\infty = \frac{2\kappa}{b}$$ (2.4.8)

and we may speak of the "mass" of a soliton, which is given as

$$M = m(y_{-\infty} - y_\infty) = \frac{2\kappa}{b} m.$$ (2.4.9)

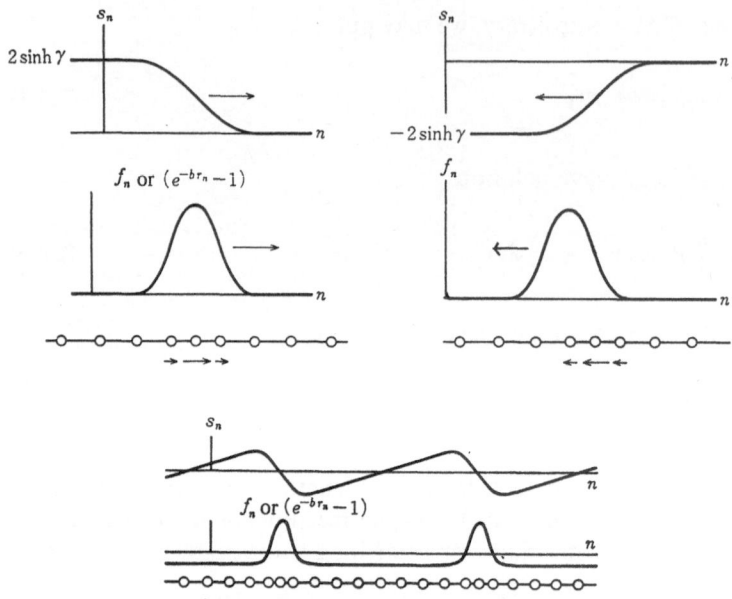

Fig. 2.4. Relation between a soliton and a cnoidal wave

Problem 2.8. Show that a soliton (2.4.5) has the momentum

$$P = Mc = \frac{2m}{b}\beta,$$

where c is the speed of the soliton given by (2.4.6).

Hint: $P = \sum m\dot{y}_n = \sum (s_{n-1} - s_n) = s_{-\infty} - s_{\infty}$.

Problem 2.9. The energy of the lattice can be written as

$$E = \sum_n \left[\frac{a}{b}(e^{-br_n} - 1) + ar_n \right] + \sum_n \frac{m}{2}\dot{y}_n^2,$$

Show that a soliton (2.4.5) has the energy

$$E = \frac{2a}{b}(\sinh\kappa \cosh\kappa - \kappa)$$

where it is assumed that there is no rarefaction or compression at infinity.

Problem 2.10. Show that we have no solution of the type (the dark soliton type)

$$e^{-br_n} - 1 = \gamma^2 - \alpha^2 \operatorname{sech}^2(\kappa n - \beta t)$$

for the lattice with exponential interaction.

Problem 2.11. Show that we have a divergent solution of the type

$$e^{-br_*} - 1 = - \alpha^2 \operatorname{cosech}^2 (\kappa n - \beta t)$$

for the lattice with exponential interaction

Remark: We have this solution from a soliton solution (2.4.5) when the variable is displaced by $i\pi/2$ $(i = \sqrt{-1})$.

Problem 2.12. Show that a soliton (2.4.5) can be derived from a cnoidal wave (2.3.1, 2) by taking the limit of large waveleneth, $\lambda \to \infty$ and large modulus, $k \to 1$.

Remark: In the limit as $k \to 1$, we have $\operatorname{sn}u \to \tanh u$, $\operatorname{dn}u \to \operatorname{cn}u \to \operatorname{sech}u$. Thus, hyperbolic functions can be included in elliptic functions. Their addition theorems are nonlinear.

2.5 Two-Soliton Solutions [2.3, 4]

Generalizing soliton solutions (2.4.2) or (2.4.4), we put

$$S_n = \frac{m}{b} \ln \left[\cosh (\mu_1 n - \gamma_1 t) + A \cosh (\mu_2 n - \gamma_2 t + \delta)\right] \tag{2.5.1}$$

or

$$S_n = \frac{m}{b} \ln \left(1 + A_1 e^{2(\kappa_1 n - \beta_1 t)} + A_2 e^{2(\kappa_2 n - \beta_2 t)} + e^{2[(\kappa_1 + \kappa_2)n - (\beta_1 + \beta_2)t]}\right) \tag{2.5.2}$$

where the parameters A, A_1, A_2, β_1, and β_2 are assumed to be functions of μ_1 and μ_2, or κ_1 and κ_2. The relations between these are, for example,

$$\left.\begin{array}{ccc} 2\kappa_1 = \mu_1 + \mu_2, & 2\beta_1 = \gamma_1 + \gamma_2, & A_1 = Ae^{\delta} \\ 2\kappa_2 = \mu_1 - \mu_2, & 2\beta_2 = \gamma_1 - \gamma_2, & A_2 = Ae^{-\delta} \end{array}\right\}. \tag{2.5.3}$$

We may take either (2.5.1) or (2.5.2). But, in what follows, we take (2.5.2). Inserting this into the equations of motion (2.2.13), we see that (2.2.13) is satisfied by

$$\beta_1^2 = \frac{ab}{m} \sinh^2 \kappa_1, \qquad \beta_2^2 = \frac{ab}{m} \sinh^2 \kappa^2$$

$$A_1 A_2 = \frac{(m/ab)(\beta_1 + \beta_2)^2 - \sinh^2 (\kappa_1 + \kappa_2)}{\sinh^2 (\kappa_1 - \kappa_2) - (m/ab)(\beta_1 - \beta_2)^2} \tag{2.5.4}$$

Due to the possible combinations of signs of β_1 and β_2 we have four cases. However, motions in which the right and the left are interchanged can be considered the same. Thus we have two independent cases, and we may assume $\kappa_1 > \kappa_2 > 0$ without losing generality.

I) The case where $\beta_1\beta_2 > 0$. We write

$$\beta_1 = \sqrt{\frac{ab}{m}} \sinh \kappa_1, \qquad \beta_2 = \sqrt{\frac{ab}{m}} \sinh \kappa_2$$

$$A_1A_2 = \left(\frac{\sinh [(\kappa_1 + \kappa_2)/2]}{\sinh [(\kappa_1 - \kappa_2)/2]}\right)^2 \tag{2.5.5}$$

To see the waveform, first consider the region where

$$\varphi_1 \equiv \kappa_1 n - \beta_1 t \simeq 0. \tag{2.5.6}$$

Then we have

$$\varphi_2 \equiv \kappa_2 n - \beta_2 t = \frac{\kappa_2}{\kappa_1} \varphi_1 + \epsilon t \tag{2.5.7}$$

where

$$\epsilon = \frac{\kappa_2}{\kappa_1} \beta_1 - \beta_2 .$$

Because we are assuming $\kappa_1 > \kappa_2 > 0$, we have $\beta_1/\beta_2 = \sinh \kappa_1/\sinh \kappa_2 > \kappa_1/\kappa_2$. Therefore $\epsilon > 0$, and in this region

$$S_n \simeq \frac{m}{b} \ln (1 + A_1 e^{2(\kappa_1 n - \beta_1 t)}) \qquad (t \to -\infty)$$

$$S_n \simeq \frac{m}{b} \ln [(A_2 + e^{2(\kappa_1 n - \beta_1 t)}) e^{2(\kappa_2 n - \beta_2 t)}] \qquad (t \to +\infty) \tag{2.5.8}$$

Similarly, for the region $\varphi_2 \equiv \kappa_2 n - \beta_2 t \simeq 0$, we have

$$S_n \simeq \frac{m}{b} \ln [(A_1 + e^{2(\kappa_2 n - \beta_2 t)}) e^{2(\kappa_1 n - \beta_1 t)}] \qquad (t \to -\infty)$$

$$S_n \simeq \frac{m}{b} \ln (1 + A_2 e^{2(\kappa_2 n - \beta_2 t)}) \qquad (t \to +\infty) \tag{2.5.9}$$

Therefore, asymptotically we have two solitons which may be written as

$$e^{-br_n} - 1 = \begin{cases} \dfrac{m}{ab} \beta_1^2 \operatorname{sech}^2 (\kappa_1 n - \beta_1 t + \delta^{(\pm)}) \\[2ex] \dfrac{m}{ab} \beta_2^2 \operatorname{sech}^2 (\kappa_2 n - \beta_2 t - \delta^{(\pm)}) \end{cases} \qquad (t \to \pm \infty), \tag{2.5.10}$$

where the phase shifts are given by

$$\delta^{(-)} = \frac{1}{2} \ln A_1, \qquad \delta^{(+)} = -\frac{1}{2} \ln A_2 . \qquad (2.5.11)$$

In this case two solitons propagate in the same direction, and one (κ_1) of them overtakes the other (κ_2). Soliton masses M_1 and M_2 are proportional to κ_1 and κ_2, and the center of mass moves with a constant velocity (cf. Fig. 2.5).

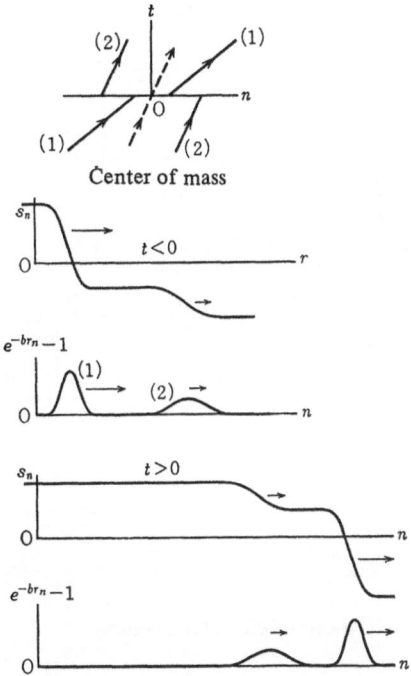

Fig. 2.5. Overtaking collision of two solitons

II) The case where $\beta_1\beta_2 < 0$. If we let $\beta_2 < 0$, we have

$$\beta_1 = \sqrt{\frac{ab}{m}} \sinh \kappa_1, \qquad \beta_2 = -\sqrt{\frac{ab}{m}} \sinh \kappa_2$$

$$A_1A_2 = \left| \frac{\cosh\left[(\kappa_1 + \kappa_2)/2\right]}{\cosh\left[(\kappa_1 - \kappa_2)/2\right]} \right|^2 \qquad (2.5.12)$$

Calculation similar to the above leads to the asymptotic form

$$e^{-br_*} - 1 = \begin{cases} \dfrac{m}{ab} \beta_1^2 \operatorname{sech}^2 (\kappa_1 n - \beta_1 t + \delta^{(\pm)}) \\[2mm] \dfrac{m}{ab} \beta_2^2 \operatorname{sech}^2 (\kappa_2 n + |\beta_2| t - \delta^{(\pm)}) \end{cases} \qquad (t \to \pm \infty), \qquad (2.5.13)$$

where

$$\delta^{(-)} = \frac{1}{2} \ln A_1, \qquad \delta^{(+)} = -\frac{1}{2} \ln A_2 .$$
(2.5.14)

In this case two solitons propagate in the opposite directions, and the center of mass moves with a constant velocity.

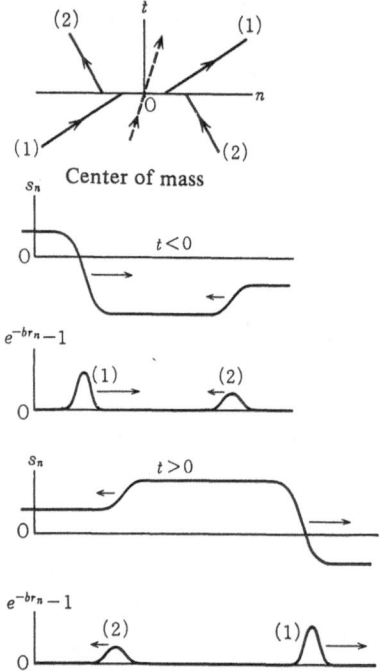

Center of mass

Fig. 2.6. Head-on collision of two solitons

It was numerically revealed by *Saito* et al. [2.5] that motion involving three solitons is stable, and meantime, *Hirota* [2.6] derived multisoliton solutions involving more than three solitons by extending (2.5.2). Since multisoliton solutions can be derived in a more systematic way, we return to this problem in the next chapter. Equation (2.5.2) consists of two exponential terms and their product, and the product term includes interaction of two solitons. Thus a multisoliton solution may be considered to be composed of essentially binary collisions.

Problem 2.13. Evaluate the maximum height of $\exp(-br_n) - 1$ in a head-on collision of solitons of the same height.

2.6 Hard-Sphere Limit

When the repulsive force is infinitely sharp and the attractive force is not present, a nonlinear lattice reduces to a system of hard spheres (in a one-dimensional system, to hard rods). In this case, each collision occurs independently, and it means only velocity exchange since we are thinking of an equal mass system. Therefore the $t \sim x$ (x stands for the position of each particle) diagram consists of straight lines. In other words, any diagram of straight lines represents a possible motion if it satisfies the condition of velocity exchange. So the motion in a one-dimensional system is very easy to understand when we consider the hard-sphere limit.

Some of the characteristic features of a nonlinear lattice can be elucidated by considering motion in a hard-sphere system. However, there are other features lost in this limit. For example, in a soliton the motion of particles is well organized and correlated. In a hard-sphere system, if we have parallel lines of several neighboring particles in the $t \sim x$ diagram the motion propagates as a pulse but it cannot be considered as a soliton.

When we give velocity to only one particle in a hard-sphere system, the resulting motion will correspond to a soliton in a nonlinear lattice. We may derive the motion from the analytic solution (2.4.1) or (2.4.5) by taking the limit of $b \to \infty$, and $a \to 0$. Intuitively, $s_n \sim \tanh x$ in (2.4.1) should become a step function, which means that $\kappa \to \infty$. The velocity of each particle is $\dot{y}_n = (s_{n-1} - s_n)/m$ by (1.4.7), which will surely represent the motion of only one particle when s_n becomes a step function. The velocity is arbitrary because it depends on the way we approach the limit $a \to 0$ and $b \to \infty$.

Let us see the hard-sphere limit of the case of a cnoidal wave [2.1a, 2]. The $t \sim x$ diagram becomes lucid when we keep the position of the minimum of the interaction potential by applying a tension $f < 0$ to the lattice, and take the limit of $a' = a + f \to 0$ and $b \to \infty$ simultaneously, to keep

$$\sigma = -\sigma' = \frac{1}{b} \ln \frac{a}{a+f} \tag{2.6.1}$$

finite. Then the interaction potential takes the form of (2.2.7). From (2.1.9) we see that $K(k) \to \infty$ (or modulus $k \to 1$) in the limit $b \to \infty$, when we keep the frequency the same, and finite motion in the system. On the other hand, Fourier expansion of the Z function is given by

$$\left. \begin{aligned} KZ(2Kx) &= 2\pi \sum_{l=1}^{\infty} \frac{q^l}{1 - q^{2l}} \sin 2\pi l x \\ q &= e^{-\pi K'/K} \end{aligned} \right\} \tag{2.6.2}$$

Taking the limit $k \to 1$, we have

$$q \simeq 1 - \varepsilon, \qquad \varepsilon = \pi^2/2K \ll 1 \qquad (2.6.3)$$

and if we keep the limiting value

$$\lim_{\substack{k \to 1 \\ b \to \infty}} \frac{K\nu}{b/m} = C \qquad (2.6.4)$$

finite, then we have the hard-sphere limit of the cnoidal wave in the form

$$\begin{aligned}
s_n &= \frac{4}{\pi} C \sum_{l=1}^{\infty} \frac{1}{l} \sin 2\pi l \left(\nu t - \frac{n}{\lambda}\right), \\
&= Cg(\nu t - n/\lambda)
\end{aligned} \qquad (2.6.5)$$

where, we note that

$$g(x) = \begin{cases} 1 - 2x & (0 < x < 1) \\ 2|x| - 1 & (-1 < x < 0). \end{cases} \qquad (2.6.6)$$

Thus the particle velocity $\dot{y}_n = (s_{n-1} - s_n)/m$ turns out to be as shown in Figs. 2.7 and 2.8.

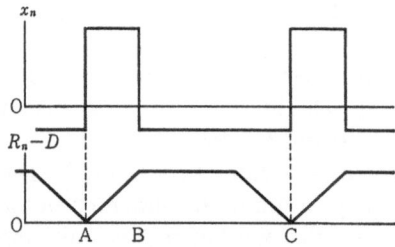

Fig. 2.7.

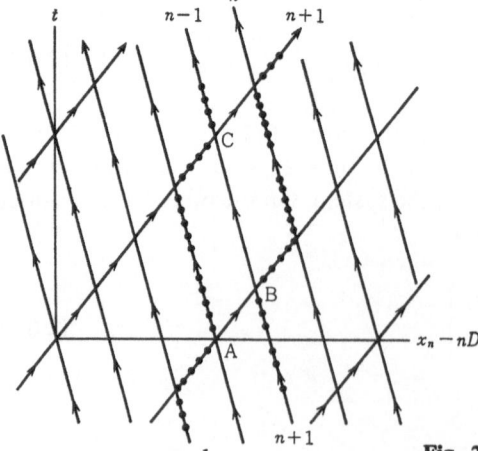

Fig. 2.8.   Hard-sphere limit of a cnoidal wave

2.7 Continuum Approximation and Recurrence Time

The continuum approximation is obtained by considering a wave whose wavelength is very long compared with the lattice spacing. We shall briefly describe a method of approximation. For this purpose it is convenient to introduce a shifting operator defined by [2.4]

$$e^{\pm \partial/\partial n} f(n) = f(n \pm 1) \,. \tag{2.7.1}$$

The equation of motion with respect to $r(n, t) = r_n(t)$ for the lattice with exponential interaction can be written as

$$m \frac{\partial^2 r}{\partial t^2} + a \left[2 \sinh \left(\frac{1}{2} \frac{\partial}{\partial n} \right) \right]^2 e^{-br} = 0 \,. \tag{2.7.2}$$

This equation can be rewritten, neglecting higher derivatives and higher powers of r, as

$$\left(\sqrt{\frac{m}{ab}} \frac{\partial}{\partial t} - 2\sinh \frac{1}{2} \frac{\partial}{\partial n} + \frac{b}{2} \frac{\partial}{\partial n} r \right) \left(\sqrt{\frac{m}{ab}} \frac{\partial}{\partial t} + 2 \sinh \frac{1}{2} \frac{\partial}{\partial n} - \frac{b}{2} r \frac{\partial}{\partial n} \right) r = 0 \,. \tag{2.7.3}$$

Therefore for a wave propagating to the right, we have

$$\left(\sqrt{\frac{m}{ab}} \frac{\partial}{\partial t} + 2 \sinh \frac{1}{2} \frac{\partial}{\partial n} - \frac{b}{2} r \frac{\partial}{\partial n} \right) r = 0 \tag{2.7.4}$$

or, formally expanding sinh up to the third-order term,

$$\sqrt{\frac{m}{ab}} \frac{\partial r}{\partial t} + \frac{\partial r}{\partial n} + \frac{1}{24} \frac{\partial^3 r}{\partial n^3} - \frac{b}{2} r \frac{\partial r}{\partial n} = 0 \,. \tag{2.7.5}$$

Introducing new variables by

$$u = \frac{1}{2} br, \qquad \tau = \sqrt{\frac{ab}{m}} t$$

$$\xi = n - \sqrt{\frac{ab}{m}} t, \qquad \delta^2 = \frac{1}{24} \tag{2.7.6}$$

we thus obtain the KdV equation

$$\frac{\partial u}{\partial \tau} - u \frac{\partial u}{\partial \xi} + \delta^2 \frac{\partial^3 u}{\partial \xi^3} = 0 \,. \tag{2.7.7}$$

For the KdV equation with respect to $u = u(\xi, \tau)$, we consider the associated eigenvalue equation

$$6\delta^2 \frac{d^2\psi^{(l)}}{d\xi^2} - (u - \lambda_l)\,\psi^{(l)} = 0\,, \tag{2.7.8}$$

where $\psi^{(l)}$ is the lth eigenfunction belonging to the eigenvalue λ_l. This equation is of the same form as the Schrödinger equation in quantum mechanics ($12\delta^2$ corresponds to $\hbar^2$). It is shown that in spite of the time evolution of $u = u(\xi, \tau)$ according to the KdV equation, the eigenvalue λ_l is independent of time. That is

$$\frac{d\lambda_l}{dt} = 0\,. \tag{2.7.9}$$

A similar statement also holds for the lattice as will be shown in the next chapter. Equation (2.7.7) admits a soliton solution of the form

$$u = u_\infty - A\,\mathrm{sech}^2\alpha(\xi - c\tau)\,, \tag{2.7.10a}$$

where u_∞ and A are constants ($A > 0$), and

$$c = -u_\infty + \frac{A}{3}\,, \qquad \alpha = \frac{1}{\delta}\sqrt{\frac{A}{12}} \tag{2.7.10b}$$

with the single eigenvalue

$$\lambda = u_\infty - \frac{A}{2}\,. \tag{2.7.11}$$

The FPU recurrence time can be discussed by using continuum approximation [2.3]. In the FPU computer experiments, initial conditions were such that the lattice was at rest at $\tau = 0$ with the displacement $y_n = B\sin(n\pi/N)$ [in this section $N + 1$ in (1.1.9) is written as N; we assume that $N \gg 1$]. Referring to linear waves, it seems plausible to assume that the wave splits into two waves with half the amplitude, which propagate in both directions along the cyclic lattice of the length $2N$. Since we can use the approximation $r = \partial y_n/\partial n$ for a smooth wave, each wave is subject to the initial condition

$$u(\tau = 0) = -\frac{bB\pi}{4N}\cos\frac{\pi}{N}\,\xi\,. \tag{2.7.12}$$

When we expand (2.7.12) around the minimum, we have

$$u(\tau = 0) = \mathrm{const.} + \frac{1}{2}\frac{bB\pi}{4N}\left(\frac{\pi}{N}\right)^2 \xi^2 \tag{2.7.13}$$

which gives a harmonic potential of the form $(\omega^2/2)\,\xi^2$ to the eigenvalue equation (2.7.8). If several solitons emerge from this initial condition, each must have the eigenvalue

$$A = \sqrt{12\delta^2 \frac{\pi^3 bB}{4N^3}} \left(l + \frac{1}{2} \right) \qquad (l = 0,1,2, \cdots) \tag{2.7.14}$$

corresponding to every level $(l + 1/2)\hbar\omega$ of a quantum harmonic oscillator. Therefore, by (2.7.10b) the speed of large solitons has the common difference

$$\Delta c = \frac{2}{3} \sqrt{12\delta^2 \frac{\pi^3 bB}{4N^3}} \tag{2.7.15}$$

and after the time interval

$$\tau_R = \frac{2N}{\Delta c} = \frac{6\sqrt{2}}{\sqrt{\pi^3}} \frac{N^{5/2}}{\sqrt{bB}}, \tag{2.7.16}$$

the original configuration will be recovered (since small solitons are somewhat out of phase, the recurrence will be incomplete). In the time scale of the lattice the recurrence time is given as

$$t_R = \frac{\tau_R}{\sqrt{ab/m}} = \frac{3\sqrt{2} N^{3/2}}{\pi^{3/2} \sqrt{bB}} t_L, \tag{2.7.17}$$

where $t_L = 2N/\sqrt{ab/m}$ is the so-called linear period. In terms of the nonlinear parameter $|\alpha| = b/2$, we have

$$t_R = \frac{3}{\pi^{3/2}} \frac{N}{\sqrt{|\alpha| B}} t_L = \frac{0.5 N^{3/2}}{\sqrt{|\alpha| B}} t_L. \tag{2.7.18}$$

This is a little larger than the empirical value (1.1.10). This discrepancy will be due to the fact that solitons pass through each other with greater speed because of the compression of the lattice.

Problem 2.14. Show that the KdV equation (2.7.7) has a periodic solution (cnoidal wave)

$$u = u_\infty - A \operatorname{cn}^2\alpha(\xi - c\tau)$$

with (k is the modulus)

$$c = -u_\infty + \frac{A}{3} \left(2 - \frac{1}{k^2} \right), \qquad \alpha = \frac{1}{k\delta} \sqrt{\frac{A}{12}}.$$

Derive this solution from the lattice cnoidal wave (2.3.1) by taking the continuum limit.

Problem 2.15. Show that

$$f = \operatorname{sech}^n(x - t)$$

satisfies

$$n^2 f_t + (n+1)(n+2) f^{2/n} f_x + f_{xxx} = 0 ,$$

where suffixes mean derivatives (for example, $f_{xxx} = \partial^3 f/\partial x^3$).

2.8 Applications and Extensions

It is easily shown that a linear lattice is equivalent to a ladder circuit (Fig. 2.9) composed of inductances L and capacitors C. Similarly a one-dimensional non-linear lattice is equivalent to a ladder circuit with nonlinear L or C. To show this, let I_n denote the current, Q_n the charge of the capacitor, Φ_n the flux in the inductance, and write the equations for the circuit as

$$\frac{dQ_n}{dt} = I_{n-1} - I_n, \qquad \frac{d\Phi_n}{dt} = V_n - V_{n+1} . \tag{2.8.1}$$

We assume that inductances are linear with constant inductance L

$$\Phi_n = LI_n . \tag{2.8.2}$$

On the other hand, let the capacitor be nonlinear with the voltage dependence

$$Q_n = Cv_0 \ln(1 + V_n/v_0) , \tag{2.8.3}$$

where C and v_0 are constants. Then (2.8.1) yields

$$Cv_0 \frac{d}{dt} \ln(1 + V_n/v_0) = I_{n-1} - I_n$$

$$L \frac{d}{dt} I_n = V_n - V_{n+1} \tag{2.8.4}$$

and we have therefore

$$CLv_0 \frac{d^2}{dt^2} \ln(1 + V_n/v_0) = V_{n-1} - 2V_n + V_{n+1} . \tag{2.8.5}$$

This is the same in form as the equations of motion (2.2.12) for a lattice with exponential interaction [2.7].

Though real capacitors may have nonlinearity different from (2.8.3), if we apply appropriate bias voltage and limit ourselves to small voltage changes we can use (2.8.3) as an approximation. By making such circuits *Hirota* and *Suzuki* [2.7] could reproduce recurrence phenomena and collision between solitons.

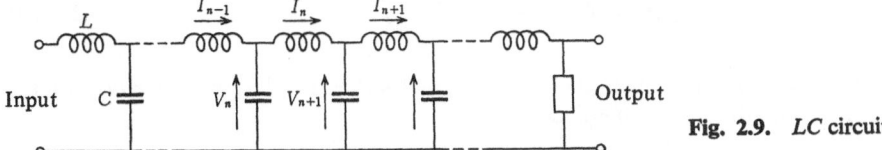

Fig. 2.9. *LC* circuit

Equation (2.8.1) may be considered as an extension of nonlinear lattice equations. However, to such an extended system mechanical interpretation will not apply in general. For example, if we assume that inductors as well as capacitors are nonlinear in such a way that

$$
\left.
\begin{aligned}
Q_n &= Cv_0 \ln\left(1 + V_n/v_0\right) \\
\Phi_n &= Li_0 \ln\left(1 + I_n/i_0\right)
\end{aligned}
\right\}
\tag{2.8.6}
$$

(C, v_0, L, i_0 are constants), then (2.8.1) gives [2.8]

$$
\left.
\begin{aligned}
\frac{d}{dt} \ln\left(z^{-1} + V_n\right) &= I_{n-1} - I_n \\
\frac{d}{dt} \ln\left(z + I_n\right) &= V_n - V_{n+1}
\end{aligned}
\right.,
\tag{2.8.7}
$$

where we have changed the scales so that t, I_n, and V_n become dimensionless by the transformations

$$
\frac{t}{\sqrt{LC}} \rightarrow t, \qquad \frac{I_n}{z^{-1}i_0} \rightarrow I_n, \qquad \frac{V_n}{zv_0} \rightarrow V_n,
\tag{2.8.8}
$$

and

$$
z = \sqrt{\frac{L}{C}\frac{i_0}{v_0}}
\tag{2.8.9}
$$

is a constant.

If $z = 1$, (2.8.7) does not change when I_n and V_n are interchanged, and it is self-dual in this sense. If $z \gg 1$, (2.8.7) reduces to the lattice with exponential interaction.

In general, (2.8.7) is altered when we replace t and I_n by $-t$ and $-I_n$. This means that, for finite z, waves propagating to the right are different in nature from waves propagating to the left. Even for this case we have multi-soliton solutions [2.8].

Nonlinear lattices can provide models for nonlinear phenomena such as wave propagation in nerve systems, chemical reactions, and certain ecological systems[2.9]. For example, we have the famous Volterra equation for ecological systems

$$\frac{dN_i}{dt} = \alpha_i N_i + \sum_j \beta_{ij} N_i N_j \qquad (i = 1, 2, \cdots) , \tag{2.8.10}$$

where α_i and β_i are constants (α_i, $\beta_i \gtrless 0$). As the simplest case we consider a chain of reactions described by

$$\frac{d}{dt} N_n = N_{n-1} N_n - N_n N_{n+1} \qquad (n = \cdots, 0, 1, 2, \cdots) . \tag{2.8.11}$$

We have a steady solution characterized by $N_{2n-1} = \bar{N}_1 = $ const. and $N_{2n} = \bar{N}_2$ $= $ const. If we put

$$\begin{aligned} N_{2n-1} &= \bar{N}_1 + n_1(n) \\ N_{2n} &= \bar{N}_2 + n_2(n) , \end{aligned} \tag{2.8.12}$$

we obtain a set of equations [2.8]

$$\begin{aligned} \frac{d}{dt} \ln [\bar{N}_1 + n_1(n)] &= n_2(n - 1) - n_2(n) \\ \frac{d}{dt} \ln [\bar{N}_2 + n_2(n)] &= n_1(n) - n_1(n + 1) \end{aligned} \tag{2.8.13}$$

which is the same as (2.8.7). We may interpret (2.8.11) as an ecological system where the nth species increases by eating the $(n - 1)$th species, and diminishes by being eaten by the $(n + 1)$th species. If we adopt this interpretation, it helps us in understanding the fact that waves propagating to the right are different, for example, in speed from waves propagating to the left.

Problem 2.16. Examine the nonlinear wave equation, which is obtained from (2.8.1) by assuming linear capacitors and nonlinear inductances of the type (2.8.6).

2.9 Poincaré Mapping

In Sect. 1.2 we discussed the behavior of a cyclic nonlinear lattice composed of three particles. As was shown by *Ford*, a three-particle cyclic lattice with cubic nonlinearity of interaction potential is equivalent to the Hénon-Heiles system, and it was shown by the method of Poincaré mapping that the trajectories in phase space become apparently stochastic when the energy exceeds a certain critical value.

    Saito [2.5] applied the method of mapping to a lattice with exponential interaction composed of two movable particles with both ends fixed, and found that its trajectories do not become erratic even when the energy is raised extremely high. *Ford* [2.10] examined in detail mapping of a three-particle cyclic lattice with exponential interaction, and obtained similar results to be described in the following.

The Hamiltonian for a three-particle cyclic lattice with exponential interaction can be written in a dimensionless form as

$$H = \frac{1}{2}(P_1^2 + P_2^2 + P_3^2) + e^{-(Q_1-Q_3)} + e^{-(Q_2-Q_1)} + e^{-(Q_3-Q_2)} - 3 \,. \qquad (2.9.1)$$

If we linearize this for a small amplitude we obtain (1.2.1). Taking up the lowest nonlinear term, we have the nonlinearity parameter of (2.2.4) as $\alpha = -1/2$ because $b = 1$ in (2.9.1). We apply the transformation (1.2.2) which diagonalizes the corresponding harmonic lattice. Then, by rescaling

$$\left. \begin{array}{ll} \zeta_1 = 2\sqrt{2}\,q_1, & \zeta_2 = 2\sqrt{2}\,q_2 \\ t \rightarrow t/\sqrt{3} & \end{array} \right\} \qquad (2.9.2)$$

we have the equations of motion

$$\begin{aligned} \ddot{q}_1 &= (4\sqrt{3})^{-1}(-e^{2q_2+2\sqrt{3}q_1} + e^{2q_2-2\sqrt{3}q_1}) \\ \ddot{q}_2 &= \frac{1}{6}e^{-4q_2} - \frac{1}{12}(e^{q_2+2\sqrt{3}q_1} + e^{2q_2-2\sqrt{3}q_1}) \end{aligned} \qquad (2.9.3)$$

The corresponding energy is

$$E = \frac{1}{2}(p_1^2 + p_2^2) + \frac{1}{24}(e^{2q_2+2\sqrt{3}q_1} + e^{2q_2-2\sqrt{3}q_1} + e^{-4q_2}) - \frac{1}{8} \,. \qquad (2.9.4)$$

For mapping we use the surface of section $(q_2 p_2)$ through $q_1 = 0$ in a four-

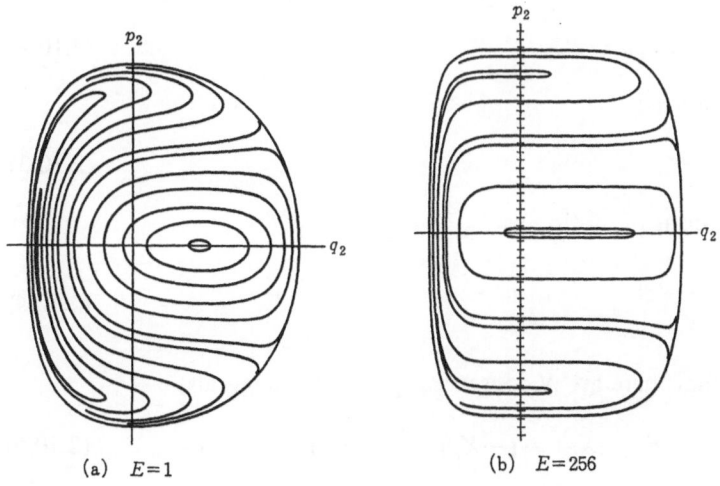

(a) $E=1$ (b) $E=256$

Fig. 2.10. Surface of section for the Hamiltonian (2.9.4) [2.10]

dimensional phase space and plot the point where trajectories cut the surface in the positive direction ($p_1 < 0$). Then we have the mapping shown in Fig. 2.10a ($E = 1$) and Fig. 2.10b ($E = 256$).

Ford examined mapping up to $E = 56000$, and it always had smooth curves on the ($q_2 p_2$) plane with no indication of stochastic behavior, showing that the trajectories were on smooth surfaces. In this case the distance (1.2.9) in phase space had linear time dependence on the average. The same was shown for a six-particle cyclic lattice with exponential interaction. Thus the numerical work strongly suggested that the lattice with exponential interaction is integrable; in other words, it admits the so-called third integral besides momentum and energy integrals.

2.10 Conserved Quantities

Stimulated by Ford's numerical work, *Hénon* and *Flaschka* independently showed analytically the integrability of the lattice with exponential interaction. Flaschka's method is discussed in the next section. First we introduce Hénon's work and related subjects [2.11].

The equations of motion for the lattice with exponential interaction can be written as

$$
\begin{aligned}
\dot{Q}_n &= P_n \\
\dot{P}_n &= C(e^{-(Q_n-Q_{n-1})} - e^{-(Q_{n+1}-Q_n)})
\end{aligned}
\tag{2.10.1}
$$

After Hénon, we have introduced a constant C, though we may put $C = 1$ in what follows. We consider a periodic lattice of N particles, so that

$$
Q_{n+N} = Q_n, \qquad P_{n+N} = P_n .
\tag{2.10.2}
$$

Further, we introduce

$$
X_n = Ce^{-(Q_{n+1}-Q_n)}
\tag{2.10.3}
$$

to write the equations of motion as

$$
\begin{aligned}
\dot{X}_n &= (P_n - P_{n+1}) X_n \\
\dot{P}_n &= X_{n-1} - X_n
\end{aligned}
\tag{2.10.4}
$$

Hénon showed that there are N conserved quantities (integrals)

$$
I_m = \sum P_{l_1} P_{l_2} \cdots P_{l_k}(-X_{j_1}) \cdots (-X_{j_l}) \qquad (m = 1, 2, \cdots, N)
\tag{2.10.5}
$$

where $\sum$ means the sum of all terms which satisfy the following conditions:

I) The indices $i_1, i_2, \ldots, i_k, j_1, j_1 + 1, \ldots, j_l, j_l + 1$ are all different (modulo N), where j's appear either explicitly or implicitly through X_j).

II) The number of these indices is m, i.e., $m = k + 2l$. Two terms differing only in the order of the factors are not considered different and, therefore, only one of them appears in the sum. For example, for $N = 3$ the integrals are

$$I_1 = P_1 + P_2 + P_3$$
$$I_2 = P_1P_2 + P_2P_3 + P_3P_1 - X_1 - X_2 - X_3 \qquad (2.10.6)$$
$$I_3 = P_1P_2P_3 - P_1X_2 - P_2X_3 - P_3X_1 .$$

To show that I_m are conserved quantities, it is convenient to write symbolically

$$P_i = [i], \qquad -X_j = [j, j + 1] \qquad (2.10.7)$$

and write (2.10.4) as

$$\frac{d}{dt}[i, i + 1] = ([i] - [i + 1])[i, i + 1] \qquad (2.10.8a)$$

$$\frac{d}{dt}[i] = -[i - 1, i] + [i, i + 1] . \qquad (2.10.8b)$$

Therefore $\dot{I}_m$ is a sum of terms similar to (2.10.5), except that one index may eventually appear twice. Below we consider all possible terms in $\dot{I}_m$, each being derived from an original term in (2.10.5).

a) A term in $\dot{I}_m$ which has no doubled index. This can only result from the derivation of a factor $[i]$ in (2.10.5), which yields $-[i - 1, i] + [i, i + 1]$, or the indices $i - 1$ and $i + 1$ by (2.10.8b). For instance, consider $i + 1$ (the same for $i - 1$). In order that $i + 1$ does not appear twice in this term, the original term should not involve $i + 1$. But, then, among the terms in I_m, we have another term where $[i]$ is replaced by $[i + 1]$, the rest being unchanged, and this term produces a derived term containing $-[i, i + 1]$ which destroys the previous one. Thus all derived terms of this kind disappear.

b) A term in $\dot{I}_m$ which has a double index, common to a factor $[i]$ and a factor $[i, i + 1]$. This can result from two original terms. (I) $[i, i + 1]$ in an original term yields $[i][i, i + 1]$ by (2.10.8a). (II) An original term with $[i][i + 1]$ is also present and yields $-[i][i, i + 1]$ through differentiation of $[i + 1]$ by (2.10.8b). Thus contributions (I) and (II) cancel each other.

c) The case of a doubled index i, common to a factor $[i - 1, i]$ and a factor $[i]$ is treated in the same way.

d) A term in $\dot{I}_m$ which has a double index i, common to a factor $[i - 1, i]$ and a factor $[i, i + 1]$. This results from an original term containing $[i - 1]$ $[i, i + 1]$ or $[i - 1, i][i + 1]$, and again derived terms cancel each other.

Thus we have

$$\dot{I}_m = 0 \quad \therefore \quad I_m = \text{const.} \tag{2.10.9}$$

That all I_m's are mutually independent is clear when we consider the case $C = 0$. In this case I_m reduces to the symmetric function $I_m = \sum P_{i_1} P_{i_2} \cdots P_{i_m}$, and the N symmetric functions I_m $(m = 1, 2, \cdots, N)$ are independent.

I_1 is simply the total momentum. I_2 is essentially the energy integral because it is related to the total energy H by

$$I_2 = \frac{1}{2} I_1^2 - H \,. \tag{2.10.10}$$

Other integrals $I_3, \ldots, I_N$ cannot be related to simple mechanical concepts.

The above conserved quantities of Hénon also apply to the lattice with fixed boundaries, which was numerically studied by *Saito* et al. [2.5]. This was pointed out independently by *Ford* and by *Toda*. To show this, we consider a periodic lattice composed of $2N + 2$ particles. Numbering the particles from $-N$ to $N + 1$, we impose the asymmetric initial condition

$$Q_{-n} = -Q_n, \quad P_{-n} = -P_n \,. \tag{2.10.11}$$

Then it is clear that this condition is maintained all the time, and it means the fixed-end condition $Q_0 = Q_{N+1} = 0$. In this case I_m's, with odd m vanish because of the symmetry, and it can be shown that I_{2N+2} reduces to a constant. Therefore we are left with N independent integrals $I_2, I_4, \ldots, I_{2N}$. Any functions of $I_1, I_2, \ldots$ are also conserved. Therefore the choice of independent conserved quantities is not unique.

The orthodox way of deriving conserved quantities will be to start with the general equation of motion for a dynamical quantity A [2.12]

$$\frac{dA}{dt} = [A, H], \tag{2.10.12}$$

where the bracket on the right-hand side means the Poisson bracket, and H is the Hamiltonian (2.10.1) $(C = 1)$ of the periodic lattice with exponential interaction, that is,

$$H = \frac{1}{2} \sum_n P_n^2 + \sum_n (e^{-(Q_{n+1}-Q_n)} - 1) \,. \tag{2.10.13}$$

We may write

$$[A, H] = \mathscr{X} A \tag{2.10.14}$$

with

$$\mathscr{X} = \sum_n P_n \frac{\partial}{\partial Q_n} - \sum e^{-(Q_{n+1}-Q_n)} \left(\frac{\partial}{\partial P_n} - \frac{\partial}{\partial P_{n+1}} \right) . \tag{2.10.15}$$

Defining

$$\rho = \sum e^{-(Q_{n+1}-Q_n)} \frac{\partial^2}{\partial P_n \partial P_{n+1}} \qquad (2.10.16)$$

we can show, after some calculation, that [2.12]

$$\mathscr{H} e^{-\rho} \prod_{i=1}^{N} P_i = 0 . \qquad (2.10.17)$$

Therefore $e^{-\rho} \prod_{i=1}^{N} P_j$ is an integral of motion. Further it can be shown that

$$\mathscr{H} \left(\sum_j \frac{\partial}{\partial P_j}\right)^n e^{-\rho} \prod_{i=1}^{N} P_i = 0 . \qquad (2.10.18)$$

The function appearing here is nothing but Hénon's integral. Indeed, by inspection, we see that Hénon's integral ($C = 1$) can be written as

$$I_{N-n} = \left(\sum_j \frac{\partial}{\partial P_j}\right)^n e^{-\rho} \prod_{i=1}^{N} P_i . \qquad (2.10.19)$$

The conserved quantities are in involution with the Hamiltonian ($[H, I_m] = 0$), and mutually in involution ($[I_n, I_m] = 0$). Further, motions which are derived respectively by taking the conserved quantities I_m as their Hamiltonians (though mechanical interpretation is impossible except for I_1 and I_2) are not mutually contradictory.

Problem 2.17. Examine the integrals of Hénon's type for the lattice $N = 3$ with fixed-ends $n = 0, n = 3$.

Problem 2.18. Verify that (2.10.19, 6) coincide for a periodic lattice.

Problem 2.19. Multiplying P_n^α ($\alpha = 0,1,2, \ldots$) to the second of the equations of motion (2.10.4) for a lattice, and summing up, derive the equations

$$J_1 = \sum_n P_n = \text{const.}$$

$$J_2 = \sum_n \left(\frac{1}{2} P_n^2 + X_n\right) = \text{const.}$$

$$J_3 = \sum_n \left[\frac{1}{3} P_n^3 + P_n(X_n + X_{n+1})\right] = \text{const.}$$

for a periodic system or a lattice at rest at infinity.

Remark: We have the relation $I_2 = J_1^2 - 2J_2$ to Hénon's integral I_2, and so forth. In a similar way, by rewriting $dp_n^4/dt, dp_n^5/dt, \ldots$, we can derive higher order conserved quantities $J_4, J_5, \ldots$. But the calculation is very tedious.

3. The Spectrum and Construction of Solutions

In the preceding section we have seen that the lattice with exponential interaction admits many particular solutions, and the cyclic lattice has as many conserved quantities as the number of particles in the lattice. Thus we have obtained fundamental knowledge of nonlinear wave propagation. In this chapter, the equations of motion are written in matrix formalism, conserved quantities are derived from them, and it is shown that the initial value problem can be solved exactly for a infinite lattice. Further, we consider equations of motion $dL/dt = BL - LB$ with generalized matrices B, and relation of the system discussed by *Kac* and *Moerbeke* to the lattice with exponential interaction. It is then shown that we can derive certain other solutions from a known solution (the Bäcklund transformation) by the use of a generating function for canonical transformation. Further, the methods, including that by *Moser,* of integration for a lattice of finite number of particles with exponential interaction are discussed. Finally, we study continuum approximation for the above subjects, deriving formulas for the Korteweg-de Vries equation related to a nonlinear continuous medium.

3.1 Matrix Formalism

Stimulated by Ford's numerical work which revealed the integrability of the lattice with exponential interaction, *Flaschka* [3.1] started an analytical survey of the lattice. At that time, it had been known that the eigenvalues of the Schrödinger equation associated to the KdV equation are time independent (cf. Sect. 2.7), and the inverse scattering method of solving the initial value problem for this equation had been invented. The theory was firmly formulated by *Lax* [3.2]. On the other hand, a method of discretization of the Schrödinger equation was established by *Kac* and *Case* [3.3]. These results are all introduced in the analytic study of the lattice with exponential interaction.

We write the equations of motion for the lattice as

$$\left.\begin{aligned} \dot{Q}_n &= P_n \\ \dot{P}_n &= e^{-(Q_n - Q_{n-1})} - e^{-(Q_{n+1} - Q_n)} \end{aligned}\right\} \tag{3.1.1}$$

and further write

$$a_n = \frac{1}{2} e^{-(Q_{n+1}-Q_n)/2}$$
$$b_n = \frac{1}{2} P_n$$

$$\left. \right\} \qquad (3.1.2)$$

The numbering n and the signs of a_n and b_n here are somewhat different from those of the original paper by Flaschka. These also differ in the works by Kac and others. In this book we use simple notation which provides us with a smooth transition in the continuum approximation to the KdV equation for waves propagating to the right. From (3.1.1) we see that the equations of motion for a_n and b_n are

$$\dot{a}_n = a_n(b_n - b_{n+1})$$
$$\dot{b}_n = 2(a_{n-1}^2 - a_n^2)$$

$$\left. \right\} \qquad (3.1.3)$$

We note that these equations are not altered when we change the sign of a_n.

We consider a periodic lattice of N particles, so that

$$a_{n+N} = a_n, \qquad b_{n+N} = b_n . \qquad (3.1.4)$$

Then we introduce $N \times N$ matrices L and B by

$$L = \begin{pmatrix}
b_1 & a_1 & & & & & & a_N \\
a_1 & b_2 & & & & & & \\
& & \ddots & & & & 0 & \\
& & & b_{n-1} & a_{n-1} & & & \\
& & & a_{n-1} & b_n & a_n & & \\
& & & & a_n & b_{n+1} & & \\
& 0 & & & & & \ddots & \\
& & & & & & b_{N-1} & a_{N-1} \\
a_N & & & & & & a_{N-1} & b_N
\end{pmatrix} \qquad (3.1.5)$$

$$B = \begin{pmatrix}
0 & -a_1 & & & & & & a_N \\
a_1 & 0 & & & & & & \\
& & \ddots & & & & 0 & \\
& & & 0 & -a_{n-1} & & & \\
& & & a_{n-1} & 0 & -a_n & & \\
& & & & a_n & 0 & & \\
& & & & & & \ddots & \\
& & & & & & 0 & -a_{N-1} \\
-a_N & & & & & & a_{N-1} & 0
\end{pmatrix} \qquad (3.1.6)$$

to write (3.1.3) as a matrix equation [3.1]

$$\frac{dL}{dt} = BL - LB .$$ (3.1.7)

Lax's formalism is to write a time evolution equation in the form of (3.1.7). Here it is essential that B is antisymmetric.

Since B is antisymmetric the matrix U defined by

$$\frac{dU}{dt} = BU, \qquad U(0) = 1$$ (3.1.8)

is unitary; that is

$$\frac{dU^{-1}}{dt} = - U^{-1}B$$ (3.1.9)

$$UU^{-1} = U^{-1}U = 1 .$$ (3.1.10)

Therefore we have

$$\frac{d}{dt}(U^{-1}LU) = 0$$ (3.1.11)

so that $U^{-1}LU$ is time independent, and

$$L(t) = U(t)L(0)U(t)^{-1} .$$ (3.1.12)

Thus $L(t)$ and $L(0)$ are unitary equivalent.

Let $\lambda(t)$ and $\varphi(t)$ ($N \times 1$ matrix) denote the eigenvalue and the eigenfunction of $L(t)$; then at $t = 0$,

$$L(0)\,\varphi(0) = \lambda(0)\,\varphi(0)$$ (3.1.13)

and using (3.1.12) we have

$$L(t)U(t)\,\varphi(0) = \lambda(0)U(t)\,\varphi(0) .$$ (3.1.14)

Comparing this with the equation

$$L(t)\varphi(t) = \lambda(t)\varphi(t)$$ (3.1.15)

at time t, we see that

$$\varphi(t) = U(t)\varphi(0)$$ (3.1.16)

or

$$\frac{d\varphi}{dt} = B\varphi \qquad\qquad (3.1.17)$$

and

$$\lambda(t) = \lambda(0) = \lambda . \qquad\qquad (3.1.18)$$

Therefore the eigenvalue is independent of time. Further, since all the elements of the matrix L are real, all the eigenvalues λ are also real.

Thus motion in the lattice conserves its spectrum λ (the isospectral deformation). By (3.1.15), the eigenvalues are determined by the determinant equation

$$\det (\lambda I - L) = 0 . \qquad\qquad (3.1.19)$$

As L is an $N \times N$ matrix, there are N eigenvalues $\lambda_1, \lambda_2, \ldots, \lambda_N$. Expanding (3.1.19) we have a set of N equations

$$\lambda_j^N + c_1 \lambda_j^{N-1} + c_2 \lambda_j^{N-2} + \cdots + c_{N-1}\lambda_j + c_N = 0 \,(j = 1,2, \cdots, N) \qquad (3.1.20)$$

where c_j's are functions of a_n and b_n. If we solve these simultaneous equations for c_j, we have

$$c_j(\{a_n\}, \{b_n\}) = c_j(\lambda_1, \lambda_2, \cdots, \lambda_N) . \qquad\qquad (3.1.21)$$

Since λ_j's are conserved, c_j's are also conserved quantities. It is easily verified that c_j's are essentially the same as Hénon's conserved quantities I_j.

As we see directly from (3.1.7)

$$\hat{J}_p = \mathrm{tr} L^p = \sum_{j=1}^{N} \lambda_p^j \qquad (p = 1,2, \cdots, N) \qquad\qquad (3.1.22)$$

(tr means the diagonal sum) are conserved quantities, and I_n can be written in terms of $\hat{J}_p$.

Problem 3.1. Let $\varphi_l(t)$ be the l-th component of $\varphi(t)$, and show that

$$|\varphi(t)|^2 = \sum_{l=1}^{N} |\varphi_l(t)|^2 = |\varphi(0)|^2 ,$$

namely that the "norm" is conserved.

Problem 3.2. Rewrite (3.1.19) as

$$\det (2\lambda I - 2L) = (2\lambda)^N + I_1(2\lambda)^{N-1} + \cdots + I_N = 0$$

and verify for $N = 3$ that I_n's are Hénon's conserved quantities.

Problem 3.3. Verify the following equations:

the total momentum $P = \sum_n P_n = \text{tr} L = 2 \sum_n \lambda_n$

The total energy $H = \sum_n \frac{1}{2} P_n^2 + \sum_n e^{-(Q_{n+1}-Q_n)} = 2 \sum_n \lambda_n^2$

$$\frac{1}{2} \hat{J}_1 = I_1 = P, \qquad \frac{1}{2} \hat{J}_2 = \frac{1}{2} I_1^2 - I_2 = H.$$

3.2 Infinite Lattice

The equation of motion (3.1.7) is equivalent to (3.1.15) and (3.1.17), namely to

$$L\varphi = \lambda\varphi, \qquad \lambda = \text{const.} \tag{3.2.1}$$

$$\frac{d\varphi}{dt} = B\varphi. \tag{3.2.2a}$$

To show this, it is only necessary to differentiate (3.2.1) with respect to t to obtain

$$\frac{dL}{dt}\varphi + LB\varphi = \lambda B\varphi = BL\varphi \tag{3.2.2b}$$

or

$$\left[\frac{dL}{dt} - (BL - LB)\right]\varphi = 0. \tag{3.2.2c}$$

Thus we have (3.1.7) [for a generalization of (3.2.2a) cf. Sect. 4.6].

In the above argument we may add to B an arbitrary imaginary constant $i\omega_0$ times a unit matrix. Then B becomes antihermitian, $B^+ = -B$, and U and φ are modified by a factor $\exp(i\omega_0 t)$.

As we have seen in the preceding section, for $\lambda = \text{const.}$, B must be anti-symmetric (or antihermitian). Since φ plays a similar role as the wave function in quantum mechanics, let us call φ the wave function hereafter. Though there exists a finite norm of φ for a cyclic lattice, we may not have the norm for an infinite lattice.

Now, in several of the following sections, we consider a lattice infinitely long in both directions. Formulas in the preceding section apply if we change the numbering $n = 1 \sim N$ for L and B by $n = -N/2 \sim N/2$. Then we may take the limit as $N \to \infty$. The elements at the upper right and the lower left of (3.1.5, 6) are ignored. Thus equations of motion (3.2.1, 2) for an infinite lattice take the form [3.4]

$$(L\varphi)_n \equiv a_{n-1}\varphi(n-1) + b_n\varphi(n) + a_n\varphi(n+1) = \lambda\varphi(n) \tag{3.2.3}$$

$$\lambda = \text{const.} \qquad (n = \cdots, 1, 2, \cdots)$$

$$\frac{d\varphi(n)}{dt} = (B\varphi)_n = a_{n-1}\varphi(n-1) - a_n\varphi(n+1) .$$

(3.2.4)

We may add a term $i\omega_0 \; \varphi(n)$ (ω_0 is arbitrary) to the right-hand side of (3.2.4).

For a while we consider motion which is confined in some finite region of the lattice, assuming no motion in the distance. Therefore for $|n| \gg 1$, we have $Q_{n+1} - Q_n = 0$, $P_n = 0$, and

$$a_n = \frac{1}{2}, \qquad b_n = 0 \qquad (|n| \gg 1)$$

(3.2.5)

so that (3.2.3, 4) reduce to

$$\frac{1}{2} \left[\varphi(n-1) + \varphi(n+1) \right] = \lambda\varphi(n)$$

(3.2.6)

$$\frac{d\varphi(n)}{dt} = \frac{1}{2} \left[\varphi(n-1) - \varphi(n+1) \right] .$$

(3.2.7)

Thus we see that the asymptotic form of the wave function is, omitting a constant factor, given as

$$\varphi(n) = e^{\mp i\omega t} z^{\pm n} \qquad (z = e^{ik}) ,$$

(3.2.8)

where

$$\lambda = \frac{1}{2} \left(z + z^{-1} \right) = \cos k$$

(3.2.9)

$$\omega = \frac{1}{2i} \left(z - z^{-1} \right) = \sin k$$

(3.2.10)

and since n is an integer, we may limit the range of k as

$$0 \leq k \leq \pi \qquad \therefore \quad -1 \leq \lambda \leq 1 .$$

(3.2.11)

If we modified B by $i\omega_0$, $\varphi(n)$ in (3.2.8) should be multiplied by a factor $\exp(i\omega_0 t)$. For $|\lambda| > 1$, the asymptotic form for $n \to +\infty$ is similarly given as

$$\varphi(n) = e^{\beta t} z_1^n \qquad (|z_1| < 1) ,$$

(3.2.12)

where

$$\lambda = \frac{z_1 + z_1^{-1}}{2}$$

(3.2.13)

$$\beta = \frac{z_1^{-1} - z_1}{2} .$$

(3.2.14)

Equation (3.2.12) for φ is an asymptotic form of the wave function of a certain bound state with $|z_1| < 1$, so that $\varphi \to 0$ as $n \to +\infty$. Since we are assuming B and L to be real matrices, we may take real z_1 and β. When the bound state is moving to the right we have $\beta > 0$ $(z_1 > 0)$, and when it is moving to the left we have $\beta < 0$ $(z_1 < 0)$.

Returning to (3.2.3), we see that the wave function $\varphi(r)$ is scattered or bound by the "potential" $a_n - 1/2$ and b_n of the motion in the lattice where $a_n \neq 1/2$ or $b_n \neq 0$. We call it "scattering" in general. As a matter of fact, (3.2.3) is a difference equation of the second rank and, therefore, reduces to the Schrödinger equation of the form $[-\hbar^2(d^2/dx^2) + u]\,\varphi = \lambda\varphi$ in the continuum approximation, where the potential u is related to $a_n - 1/2$ and b_n. If we are given the potential and ask the scattering and bound state of φ to see its asymptotic form, it is an ordinary scattering problem. Conversely, if we look for the potential when the asymptotic form of φ is given, this is the so-called inverse scattering method (I.S.M.).

Suppose we want to have detailed knowledge of P_n and Q_n or of a_n and b_n of a moving object, and illuminate it with a sort of light characterized by λ. A radar image thus obtained corresponds to the asymptotic form of φ, and the technique of predicting the movement of the object corresponds to the inverse scattering method. The wave function φ is nothing but a mathematical tool, but we may think of scattering of a realistic wave, such as x-rays, by solitons in a nonlinear substance [3.5].

As we shall see later, solitons are specified by bound states. Therefore, the spectrum of bound states ($|\lambda| > 1$) and their asymptotic form are more important in characterizing nonlinear wave propagation than the scattering of wave functions with $|\lambda| < 1$. For a periodic system, to be discussed in Chap. 5, we cannot speak of asymptotic form at infinity, and instead we make use of an auxiliary spectrum to solve initial value problems of lattice waves. Thus such methods may be compiled as the inverse spectral theory (IST) rather than as the inverse scattering method.

If, in general, we put

$$\lambda = \frac{z + z^{-1}}{2}, \tag{3.2.15}$$

then (3.2.3) becomes

$$a_{n-1}\varphi(n-1) + b_n\varphi(n) + a_n\varphi(n+1) = \frac{1}{2}(z + z^{-1})\varphi(n)\,. \tag{3.2.16}$$

When there is no motion at infinity, the solution with the asymptotic form of (3.2.8), or

$$\varphi(n) \to z^n e^{-i\omega t}\,, \qquad i\omega = \frac{z - z^{-1}}{2} \tag{3.2.17}$$

for $n \gg 1$, can be written as

$$\varphi(n) = \phi(n, z)\,e^{-i\omega t}, \tag{3.2.18}$$

where

$$\phi(n, z) = \sum_{n'=n}^{\infty} K(n, n')\,z^{n'}. \tag{3.2.19}$$

$K(n, n')$ does not depend on z; but as a_n and b_n involve t. $K(n, n')$ also depend on t.

Proof. In the distance $n \gg 1$, from (3.2.8) we have evidently $\phi(n, z) \sim z^n (K(n, n') = 0, n \neq n')$. Thus (3.2.15) holds for sufficiently large n. Assume it holds for n and $n + 1$. Then from (3.2.16) we obtain $\varphi(n - 1, z)$ and $K(n - 1, n')$'s for $n' < n - 1$ are determined. (Q.E.D.)

Inserting (3.2.18, 19) into (3.2.16) and comparing the coefficients of z^{n-1}, z^n, z^{n+1}, . . ., we obtain

$$\left.\begin{aligned}
&a_{n-1}K(n - 1, n - 1) = \frac{1}{2}\,K(n, n) \\[6pt]
&a_{n-1}K(n - 1, n) + b_n K(n, n) = \frac{1}{2}\,K(n, n + 1) \\[6pt]
&a_{n-1}K(n - 1, n + 1) = a_n K(n + 1, n + 1) + b_n K(n, n + 1) \\[4pt]
&\quad = \frac{1}{2}\,[K(n, n) + K(n, n + 2)] \\[6pt]
&a_{n-1}K(n - 1, n + 2) + a_n K(n + 1, n + 2) + b_n K(n, n + 2) \\[4pt]
&\quad = \frac{1}{2}\,[K(n, n + 1) + K(n, n + 3)] \\[4pt]
&\cdots\cdots\cdots\cdots\cdots \\[4pt]
&a_{n-1}K(n - 1, n + m) + a_n K(n + 1, n + m) + b_n K(n, n + m) \\[4pt]
&\quad = \frac{1}{2}\,[K(n, n + n + m - 1) + K(n, n + m + 1)] \\[4pt]
&\cdots\cdots\cdots\cdots\cdots
\end{aligned}\right\} \tag{3.2.20}$$

Then we can solve for a_{n-1}, b_m. From the first two equations we have

$$\left.\begin{aligned}
a_{n-1} &= \frac{K(n, n)}{2K(n - 1, n - 1)} \\[8pt]
b_n &= \frac{K(n, n + 1)}{2K(n, n)} - a_{n-1}\frac{K(n - 1, n)}{K(n, n)}
\end{aligned}\right\} \tag{3.2.21}$$

or

$$b_n = \frac{K(n, n + 1)}{2K(n, n)} - \frac{K(n - 1, n)}{2K(n - 1, n - 1)}. \tag{3.2.22}$$

Other equations of (3.2.20) are those which $K(n, m)$ with increasing m should satisfy by turns.

Equations (3.2.21) and (3.1.2) give us

$$e^{-(Q_n - Q_{n-1})} = \left[\frac{K(n, n)}{K(n-1, n-1)} \right]^2 . \tag{3.2.23}$$

Further, $P_n = s_{n-1} - s_n$ [cf. (1.4.7) and (3.2.22) yield

$$s_n = - \frac{K(n, n+1)}{K(n, n)} , \tag{3.2.24}$$

except for an additional constant. From (2.2.14) we have also

$$e^{-(Q_{n+1} - Q_n)} - 1 = \dot{s}_n . \tag{3.2.25}$$

3.3 Scattering and Bound States [3.4, 6]

Since (3.2.16) is a difference equation of the second rank, its general solution can be expressed as a linear combination of two fundamental solutions. As fundamental solutions we choose two functions $\phi(n, z)$ and $\phi(n, z^{-1})$ with the asymptotic forms

$$\left. \begin{array}{ll} \phi(n, z) \to z^n & (n \to +\infty) \\ \phi(n, z^{-1}) \to z^{-n} & (n \to +\infty) \end{array} \right\} \tag{3.3.1}$$

[we have omitted the factor $\exp(-i\omega t)$ in (3.2.18)]. Similarly, we have solutions of (3.2.16) specified as

$$\left. \begin{array}{ll} \varphi(n, z) = \psi(n, z)\, e^{+i\omega t} \\ \psi(n, z) \to z^{-n} & (n \to -\infty) \\ \psi(n, z^{-1}) \to z^n & (n \to -\infty) \end{array} \right\} \tag{3.3.2}$$

We write these in terms of the fundamental solutions as

$$\left. \begin{array}{l} \psi(n, z) = \alpha(z)\phi(n, z^{-1}) + \beta(z)\phi(n, z) \\ \psi(n, z^{-1}) = \alpha(z^{-1})\phi(n, z) + \beta(z^{-1})\phi(n, z^{-1}) \end{array} \right\} . \tag{3.3.3}$$

Since a_n and b_n in (3.2.16) are time dependent, α and β also include time. Conversely, we write

$$\left. \begin{array}{l} \phi(n, z) = \bar{\alpha}(z)\psi(n, z^{-1}) + \bar{\beta}(z)\psi(n, z) \\ \phi(n, z^{-1}) = \bar{\alpha}(z^{-1})\psi(n, z) + \bar{\beta}(z^{-1})\psi(n, z^{-1}) \end{array} \right\} . \tag{3.3 4}$$

Therefore, compatibility conditions are

$$\left.\begin{array}{l}
\alpha(z)\bar{\alpha}(z^{-1}) + \beta(z)\bar{\beta}(z) = 1 \\
\alpha(z)\bar{\beta}(z^{-1}) + \beta(z)\bar{\alpha}(z) = 0 \\
\alpha(z^{-1})\bar{\alpha}(z) + \beta(z^{-1})\bar{\beta}(z^{-1}) = 1 \\
\alpha(z^{-1})\bar{\beta}(z) + \beta(z^{-1})\bar{\alpha}(z^{-1}) = 0
\end{array}\right\} \tag{3.3.5a}$$

$$\left.\begin{array}{l}
\bar{\alpha}(z)\alpha(z^{-1}) + \bar{\beta}(z)\beta(z) = 1 \\
\bar{\alpha}(z)\beta(z^{-1}) + \bar{\beta}(z)\alpha(z) = 0 \\
\bar{\alpha}(z^{-1})\alpha(z) + \bar{\beta}(z^{-1})\beta(z^{-1}) = 1 \\
\bar{\alpha}(z^{-1})\beta(z) + \bar{\beta}(z^{-1})\alpha(z^{-1}) = 0
\end{array}\right\} \tag{3.3.5b}$$

These are satisfied by

$$\left.\begin{array}{l}
\bar{\alpha}(z) = \alpha(z), \qquad \bar{\beta}(z) = -\beta(z^{-1}) \\
\alpha(z)\alpha(z^{-1}) = 1 + \beta(z)\beta(z^{-1})
\end{array}\right\}. \tag{3.3.6a}$$

We also have, of course, $\bar{\beta}(z^{-1}) = -\beta(z)$, etc.

For $|z| = 1$, we have

$$|\alpha(z)|^2 = 1 + |\beta(z)|^2 \tag{3.3.6b}$$

We define the "scattering function" by

$$S(n, z) = \frac{\psi(n, z)}{\alpha(z)} = \phi(n, z^{-1}) + R(z)\phi(n, z), \tag{3.3.7}$$

where

$$R(z) = \frac{\beta(z)}{\alpha(z)} \tag{3.3.8}$$

$S(n,z)$ is a solution, with the asymptotic forms

$$\left.\begin{array}{ll}
S(n, z) \to z^{-n} + R(z)z^n & (n \to +\infty) \\
S(n, z) \to z^{-n}/\alpha(z) & (n \to -\infty)
\end{array}\right\}. \tag{3.3.9}$$

We can interpret this result in the following way: as an incident wave of unit amplitude $z^{-n} = \exp(-ikn)$ propagates to the left, a reflected wave $R(z)z^n$ which proceeds to the right, and a transmitted wave $z^{-n}/\alpha(z)$ which proceeds to the left, are produced. $R(z)$ stands for the amplitude "reflection coefficient" and $1/\alpha(z)$ the amplitude "transmission coefficient"; (3.3.6b) gives the conservation of waves

$$\frac{1}{|\alpha(z)|^2} = 1 - |R(z)|^2 \tag{3.3.10}$$

We study the transmission coefficient to see its relation to the bound state. First we solve

$$\left.\begin{aligned}\psi(n, z) &= \alpha(z)\phi(n, z^{-1}) + \beta(z)\phi(n, z) \\ \psi(n + 1, z) &= \alpha(z)\phi(n + 1, z^{-1}) + \beta(z)\phi(n + 1, z)\end{aligned}\right\} \tag{3.3.11}$$

for $\alpha(z)$ by eliminating $\beta(z)$ to obtain

$$\alpha(z) = [\psi(n, z)\phi(n + 1, z) - \psi(n + 1, z)\phi(n, z)]/W(n), \tag{3.3.12}$$

where

$$W(n) = \phi(n, z^{-1})\phi(n + 1, z) - \phi(n + 1, z^{-1})\phi(n, z). \tag{3.3.13a}$$

Rewriting the last equation as

$$W(n) = \phi(n, z^{-1})[\phi(n + 1, z) - \phi(n, z)] - [\phi(n + 1, z^{-1}) - \phi(n, z^{-1})]\phi(n, z) \tag{3.3.13b}$$

we see that $W(n)$ is a discrete version of the Wronskian of $\phi(n, z^{-1})$ and $\phi(n, z)$. From the general property of the Wronskian, we may anticipate that either it is independent of n or that it has simple n dependence. To verify this assertion we eliminate b_n from

$$\left.\begin{aligned}a_{n-1}\phi(n - 1, z^{-1}) + a_n\phi(n + 1, z^{-1}) + b_n\phi(n, z^{-1}) &= \frac{z + z^{-1}}{2}\phi(n, z^{-1}) \\ a_{n-1}\phi(n - 1, z) + a_n\phi(n + 1, z) + b_n\phi(n, z) &= \frac{z + z^{-1}}{2}\phi(n, z)\end{aligned}\right\} \tag{3.3.14}$$

to obtain

$$W(n - 1)a_{n-1} = W(n)a_n. \tag{3.3.15}$$

Thus we have

$$W(n) = \frac{a_{n+1}}{a_n}W(n + 1) = \frac{a_{n+1}}{a_n}\frac{a_{n+2}}{a_{n+1}}W(n + 2) = \cdots = \frac{a_N}{a_n}W(N). \tag{3.3.16}$$

However, we have $a_N = 1/2$ for $N \to \infty$, and from (3.3.13 a)

$$W(N) \to z^{-N}z^{N+1} - z^{-(N+1)}z^N = z - z^{-1} \quad (N \to +\infty). \tag{3.3.17}$$

Therefore we have

$$W(n) = \frac{z - z^{-1}}{2a_n} \tag{3.3.18}$$

so that $a_n W(n)$ is independent of n.

Now, inserting (3.3.18) into (3.3.12) we have

$$\alpha(z) = \frac{2a_n}{z + z^{-1}} [\psi(n, z)\phi(n + 1, z) - \psi(n + 1, z)\phi(n, z)] . \tag{3.3.19}$$

As is clear from the meaning of $\alpha(z)$, the right-hand side of (3.3.19) must be independent of n (cf. Sect. 3.7).

We denote by z_j $(j = 1,2, . . .)$ the zeros of $\alpha(z)$ in the complex z-plane,

$$\alpha(z_j) = 0 . \tag{3.3.20}$$

Then by (3.3.3) we have

$$\psi(n, z_j) = \beta(z_j)\phi(n, z_j) . \tag{3.3.21}$$

Since $\varphi(n, z) \to z^{-n}$ for $n \to -\infty$, and $\phi(n, z) \to z^n$ for $n \to +\infty$, for the zeros for which

$$|z_j| < 1 , \tag{3.3.22}$$

we have

$$\psi(n, z_j) \sim \phi(n, z_j) \to 0 \quad (n \to \pm \infty) . \tag{3.3.23}$$

That is, with a normalization constant μ,

$$\zeta_j(n, z_j) = \mu\psi(n, z_j) = \mu\beta(z_j)\phi(n, z_j) \tag{3.3.24}$$

is the wave function for a bound state.

At $z = z_j$, $1/\alpha(z)$ diverges. Let us find the residue

$$\text{Res} \left\{ \frac{1}{\alpha(z)}\Big|_{z=z_j} \right\} = \frac{1}{d\alpha(z)/dz} . \tag{3.3.25}$$

In order to evaluate the denominator from (3.3.19), we first study

$$\psi'(n, z) = \frac{d\psi(n, z)}{dz} . \tag{3.3.26}$$

Differentiating the equation

$$a_{n-1}\psi(n - 1, z) + a_n\psi(n + 1, z) + b_n\psi(n, z) = \lambda\psi(n, z)$$

for (n,z), we have

$$a_{n-1}\psi'(n+1, z) + a_n\psi'(n+1, z) + b_n\psi'(n, z) = \lambda\psi'(n, z) + \frac{d\lambda}{dz}\psi(n,z).$$

$$(3.3.27)$$

If we combine this with the equation

$$a_{n-1}\phi(n-1, z) + a_n\phi'(n+1, z) + b_n\phi(n, z) = \lambda\phi(n, z)$$

for $\phi(n,z)$, we obtain

$$\frac{d\lambda}{dz}\psi(n, z)\phi(n, z) = a_{n-1}[\psi'(n-1, z)\phi(n, z) - \psi'(n, z)\phi(n-1, z)]$$

$$-a_n[\psi'(n, z)\phi(n+1, z) - \psi'(n+1, z)\phi(n, z)]. \quad (3.3.28)$$

We sum up over n and put $z = z_j$. Then by (3.3.2, 22) we have

$$\frac{d\lambda(z_j)}{dz_j}\sum_{n'=-\infty}^{n}\psi(n', z_j)\phi(n', z_j)$$

$$= -a_n[\psi'(n, z_j)\phi(n+1, z_j) - \psi'(n+1, z_j)\phi(n, z_j)]. \quad (3.3.29)$$

Here we have to note that though $\varphi'(n, z_j) = d\varphi(n,z)/dz|_{z=z_j}$ vanishes at $n \rightarrow -\infty$, it may not vanish at $n \rightarrow +\infty$.

In the same way using $\phi'(n, z)$ we obtain

$$\frac{d\lambda(z_j)}{dz_j}\sum_{n'=n+1}^{\infty}\psi(n', z_j)\phi(n', z_j)$$

$$= a_n[\phi'(n, z_j)\psi(n+1, z_j) - \phi'(n+z_j)\psi(n, z_j)]. \quad (3.3.30)$$

Here we note that though $\phi'(n, z_j) = d\phi(n, z) dz|_{z=z_j}$ vaishes at $n \rightarrow +\infty$, it may not vanish at $n \rightarrow -\infty$.

Adding (3.3.29, and 30), and using (3.3.19) we obtain

$$\frac{d\lambda(z_j)}{dz_j}\sum_{n=-\infty}^{\infty}\psi(n, z_j)\phi(n, z_j) = -a_n\frac{d}{dz}[\psi(n, z)\phi(n+1, z) - \psi(n+1, z)\phi(n, z)]_{z=z_j}$$

$$= -\frac{d}{dz}\left[\frac{z - z^{-1}}{2}\alpha(z)\right]\Big|_{z=z_j}$$

$$= -\frac{z_j - z_j^{-1}}{2}\frac{d\alpha(z)}{dz}\Big|_{z=z_j}, \quad (3.3.31)$$

where, since $\lambda = (z + z^{-1})/2$,

$$\frac{d\lambda(z_j)}{dz_j} = \frac{z_j - z_j^{-1}}{2z_j}. \quad (3.3.32)$$

Thus by (3.3.24), we have

$$\left.\frac{d\alpha(z)}{dz}\right|_{z=z_j} = -\frac{1}{z_j}\sum_{n=-\infty}^{\infty}\psi(n,z_j)\phi(n,z_j)$$

$$= -\frac{1}{z\,\mu^2\beta(z_j)}\sum_{n=-\infty}^{\infty}[\zeta(n,z_j)]^2 \ . \tag{3.3.33}$$

We normalize (3.3.24) in such a way that

$$\sum_{n=-\infty}^{\infty}[\zeta_j(n,z_j)]^2 = 1 \tag{3.3.34}$$

and write the asymptotic form as

$$\zeta_j(n,z_j) \to c_j z_j^n \qquad (n \to +\infty) \ . \tag{3.3.35}$$

Then, by (3.3.24), the normalization coefficient c_j is

$$c_j = \mu\beta(z_j) \ . \tag{3.3.36}$$

Thus we have

$$\left.\frac{d\alpha(z)}{dz}\right|_{z=z_j} = -\frac{\beta(z_j)}{z_j c_j^2} \tag{3.3.37}$$

which does not vanish. Therefore $z = z_j$ is a simple root with the residue

$$\text{Res}\left\{\left.\frac{1}{\alpha(z)}\right|_{z=z_j}\right\} = -\frac{z_j c_j^2}{\beta(z_j)}, \tag{3.3.38}$$

3.4   The Gel'fand-Levitan Equation

We multiply $z^{m-1}(m \geq n)$ to the equation (3.3.7) for the scattering function, and integrate along a circle around the origin on the complex z plane. Then we have

$$\frac{1}{2\pi i}\oint\frac{\psi(n,z)}{\alpha(z)}z^{m-1}dz = \frac{1}{2\pi i}\oint[\phi(n,z^{-1}) + R(z)\phi(n,z)]z^{m-1}dz \ . \tag{3.4.1}$$
$$\text{(lhs)} \qquad\qquad\qquad\qquad\qquad \text{(rhs)}$$

We rewrite both sides to find an equation satisfied by $K(n,n')$ of (3.2.19).
  First, for the (rhs), we recall that, for $|z| \leq 1$,

$$\phi(n,z) = \sum_{n'=n}^{\infty}K(n,n')z^{n'} \ . \tag{3.4.2}$$

Then, Cauchy's residue theorem leads to

$$\frac{1}{2\pi i} \oint \phi(n, z^{-1})\, z^{m-1} dz = \frac{1}{2\pi i} \sum_{n'=n}^{\infty} K(n, n') \oint z^{-n'+m-1} dz$$

$$= K(n, m) \qquad (m \geq n). \tag{3.4.3}$$

Noting that

$$\frac{1}{2\pi i} \oint R(z)\phi(n, z)\, z^{m-1} dz = \frac{1}{2\pi i} \sum_{n'=n}^{\infty} K(n, n') \oint R(z)\, z^{n'+m-1} dz \,,$$

we have

$$(\text{rhs}) = K(n, m) + \sum_{n'=n}^{\infty} K(n, n')\, F_{c}(n' + m) \tag{3.4.4}$$

where F_c is a contribution from the continuous spectrum of z, and is defined by

$$F_{c}(m) = \frac{1}{2\pi i} \oint R(z)\, z^{m-1} dz \,. \tag{3.4.5}$$

On the other hand, for the (lhs), the integral along the unit circle around the origin can be written as

$$(\text{lhs}) = \frac{1}{2\pi i} \oint \frac{\psi(n, z)}{\alpha(z)}\, z^{m-1} dz = I_{\alpha} + I_0 \tag{3.4.6}$$

where I_{α} are contibutions from the poles z_j $(z_j \neq 0)$, and I_0 is a contribution from a pole at $z = 0$. Since the residue at z_j is given by (3.3.38), we have, by (3.3.26),

$$I_{\alpha} = -\sum_{j} \psi(n, z_j)\, z_j^{m-1} \frac{z_j c_j^2}{\beta(z_j)} = -\sum_{j} \phi(n, z_j)\, z_j^m c_j^2$$

$$= -\sum_{j} c_j^2 \sum_{n'=n}^{\infty} K(n, n')\, z_j^{n'+m} \,. \tag{3.4.7}$$

With regard to the pole of $\psi(n, z)\, z^{m-1}$, we recall that

$$S(n, z) = \frac{\psi(n, z)}{\alpha(z)} = \phi(n, z^{-1}) + R(z)\phi(n, z) \,. \tag{3.4.8}$$

Because of the first term on the right-hand side, $S(n, z)\, z^{m-1}$ may have a pole at $z = 0$ (cf. Problem 3.1, p. 64). To verify this assertion directly, we estimate $S(n, z)$ for $z \simeq 0$ using the equation

$$a_{n-1}S(n - 1, z) + a_n S(n + 1, z) + b_n S(n, z) = \frac{z + z^{-1}}{2} S(n, z) \,. \tag{3.4.9}$$

On the right-hand side we have a factor z^{-1}. Therefore near $z \simeq 0$, we have

$$a_n S(n+1, z) \simeq \frac{1}{2z} S(n, z) \qquad (z \simeq 0) . \tag{3.4.10}$$

Thus for $z \simeq 0$ we have

$$S(n, z) \simeq 2za_n S(n+1, z) \simeq \cdots \simeq (2z)^N a_n a_{n+1} \cdots a_{n+N-1} S(n+N, z) . \tag{3.4.11}$$

By (3.2.21), $a_n = K(n+1, n+1)/2K(n, n)$ and recalling (3.3.9) we have for $N \gg 1$

$$S(N, z) \simeq z^{-N} \qquad (z \simeq 0, N \gg 1) . \tag{3.4.12}$$

Further by the asymptotic form of $\phi(n, z)$ we have

$$K(\infty, \infty) = 1 . \tag{3.4.13}$$

Using these results, we have

$$S(n, z) \simeq \frac{1}{K(n, n)} z^{-n} \qquad (z \simeq 0) . \tag{3.4.14}$$

Therefore when $-n + m - 1 = -1$, that is, when $n = m$, the pole $z = 0$ contributes I_0 of (3.4.6). Since $m \geq n$, lower terms $z^{-\nu}(\nu < n)$ in $S(n, z)$ have no contribution. Thus

$$I_0 = \frac{1}{K(n, n)} \delta(n, m) \qquad (m \geq n) , \tag{3.4.15}$$

and for the (lhs) of (3.4.1), we have

$$(\text{lhs}) = \frac{1}{K(n, n)} \delta(n, m) - \sum_{n'=n}^{\infty} K(n, n') F_b(n' + m) \qquad (m \geq n) \tag{3.4.16}$$

where F_b stands for contributions from the bound states,

$$F_b(m) = \sum_j c_j^2 z_j^m . \tag{3.4.17}$$

Therefore (3.4.1) becomes

$$\frac{1}{K(n, n)} \delta(n, m) = K(n, m) + \sum_{n'=n}^{\infty} K(n, n') F(n' + m) \qquad (m \geq n) \tag{3.4.18}$$

with

$$F(m) = \frac{1}{2\pi i} \oint R(z)\, z^{m-1} dz + \sum_j c_j^2 z_j^m\,. \tag{3.4.19}$$

This is a discrete version of the Gel'fand-Levitan (GL) equation (sometimes called the Gel'fand-Levitan-Marchenko equation). In principle, if we solve (3.4. 19) for a given kernel $F(m)$, to obtain $K(n, m)$ then the motion in the lattice is known by (3.2.19) or (3.2.20). Though (3.4.19) looks nonlinear with respect to $K(n, m)$ because of the left-hand side, it can be linearized immediately in the following way.

For the off-diagonal elements of $K(n, m)$ we write

$$\kappa(n, m) = \frac{K(n, m)}{K(n, n)} \qquad (m > n)\,. \tag{3.4.20}$$

Then, for the off-diagonal elements, GL equation (3.4.18) becomes

$$\kappa(n, m) + F(n + m) + \sum_{n'=n+1}^{\infty} \kappa(n, n')\, F(n' + m) = 0 \tag{3.4.21}$$

which is linear with respect to $\kappa(n, m)$, and can be solved in principle. When we thus obtain $\kappa(n, m)$, we have the diagonal element $K(n, n)$ from (3.4.18) for $n = m$ as

$$\frac{1}{[K(n, n)]^2} = 1 + F(2n) + \sum_{n'=n+1}^{\infty} \kappa(n, n')\, F(n' + m)\,. \tag{3.4.22}$$

3.5 The Initial Value Problem [3.3]

We have seen that the lattice with exponential interaction is described in (3.2.3, 4), and that (3.2.3) is linearized in the form of (3.4.21, 22). If we can determine, from (3.2.4), the time dependence of the kernel $F(m)$ of the discrete integral equation (3.4.21), then we can solve the initial value problem. In the preceding section we have written the asymptotic form of the scattering function as (3.3.9) disregarding the time factor given in (3.2.18). Now we shall restore the time factor $\exp(i\omega t)$ noting the incident wave, the first term z^{-n} of (3.3.9), and let $R(z, t)$ include a certain time dependence, to write the asymptotic form of (3.3.9) as

$$S(n, z, t) \to [z^{-n} + R(z, t)\, z^n]\, e^{i\omega t} \qquad (n \to +\infty)\,, \tag{3.5.1a}$$

where by (3.2.10)

$$i\omega = \frac{z - z^{-1}}{2}\,. \tag{3.5.2}$$

Since $S(n, z, t)$ is a wave function, its time dependence is given by (3.2.4) as

$$\frac{d}{dt} S(n, z, t) = a_{n-1} S(n-1, z, t) - a_n S(n+1, z, t).$$ (3.5.3)

Therefore, its asymptotic form for $n \to +\infty$ is given as

$$\frac{d}{dt} S(n, z, t) \to \frac{1}{2} [(z^{-n+1} - z^{-n-1}) + R(z, t)(z^{n-1} - z^{n+1})] e^{i\omega t}$$

$$= \frac{z - z^{-1}}{2} [z^{-n} - R(z, t) z^n] e^{i\omega t} \qquad (n \to +\infty).$$ (3.5.4)

This must be identical to the direct time derivative

$$\frac{d}{dt} S(n, z, t) \to \dot{R}(z, t) z^n e^{i\omega t} + i\omega [z^{-n} + R(z, t) z^n] e^{i\omega t} \qquad (n \to +\infty).$$ (3.5.5)

Thus we have

$$\dot{R}(z, t) = \left(\frac{z^{-1} - z}{2} - i\omega \right) R(z, t)$$
$$= (z^{-1} - z) R(z, t)$$ (3.5.6)

Therefore the time dependence of the reflection coefficient is given as

$$R(z, t) = \frac{\beta(z, t)}{\alpha(z, t)} = R(z, 0) e^{(z^{-1}-z)t}.$$ (3.5.7)

A similar calculation for

$$\psi(n, z, t) = \alpha(z, t) S(n, z, t) \to [\alpha(z, t) z^{-n} + \beta(z, t) z^n] e^{i\omega t} \qquad (n \to +\infty)$$ (3.5.1b)

yields the results

$$\left. \begin{array}{l} \alpha(z, t) = \alpha(z, 0) \\ \beta(z, t) = \beta(z, 0) e^{(z^{-1}-z)t} \end{array} \right\}.$$ (3.5.8)

For the bound state, we have only a term in the asymptotic form, and the time dependence can be attributed solely to the normalization coefficient c_j. We can thus write

$$\zeta_j(n, z_j, t) \to c_j(t) z_j^n.$$ (3.5.9)

Then, from the equation

$$\frac{d}{dt} \zeta_j(n, z_j, t) = a_{n-1}\zeta_j(n - 1, z_j, t) - a_n\zeta_j(n + 1, z_j, t) \qquad (3.5.10)$$

we have the asymptotic estimation

$$\dot{c}_j = \frac{z_j^{-1} - z_j}{2} \cdot c_j \qquad (3.5.11)$$

which results in

$$c_j(t) = c_j(0) \, e^{(z_j^{-1} - z_j)t/2} \,. \qquad (3.5.12)$$

Considering (3.5.7, 12) we have the time-dependent kernel of the GL equation as

$$F(m) = \frac{1}{2\pi i} \oint R(z, 0) \, e^{(z^{-1} - z)t} \, z^{m-1} dz + \sum_j c_j^2(0) \, e^{(z_j^{-1} - z_j)t} \, z_j^m \,. \qquad (3.5.13)$$

When the initial values $a_n(0)$ and $b_n(0)$ are given then we solve the scattering problem of the linear equation (3.2.3) under the boundary conditions (3.3.9) to obtain the initial value of the reflection coefficient $R(z, 0)$ and the initial value of the normalization coefficient $c_j(0)$. From these initial scattering data, we construct the kernel (3.5.13) and solve the linear discrete integral equation (GL equation) (3.4.18) for $\kappa(n, m)$; then we have $K(n, m)$ and equations (3.2.23–25) give the solution to the initial value problem. Thus we see that the problem of the nonlinear lattice wave propagation has been linearized, so to speak, and therefore reduced to a problem which is solvable in principle.

3.6 Soliton Solutions [3.4]

If the reflection coefficient vanishes initially $[R(z, 0) = 0]$, it remains zero all the time $[R(z, t) = 0]$. In this chapter we consider this case, and find soliton solutions which are characterized by zeros z_j of $\alpha(z)$ and normalization coefficients $c_j(0)$.

First, we discuss the case where we have only one zero, and write it as

$$z_1 = \pm\, e^{-\gamma} = \text{real} \,. \qquad (3.6.1)$$

Since we assume that $|z_1| < 1$, we have $\gamma > 0$. But we have two possibilities $z_1 > 0$ and $z_1 < 0$. The normalization coefficient belonging to z_1 can be written, by (3.5.12, 14), as

$$c_1 = c_1(0)\, e^{\beta t} \tag{3.6.2}$$

with

$$\beta = \frac{z_1^{-1} - z_1}{2} = \pm \sinh \gamma \,. \tag{3.6.3}$$

The kernel of the GL equation in this case is

$$F(m) = c_1^2 z_1^m \tag{3.6.4}$$

and the GL equation reduces to

$$\kappa(n, m) + c_1^2 z_1^{n+m} + c_1^2 z_1^m \sum_{n'=n+1}^{\infty} \kappa(n, n')\, z_1^{n'} = 0 \,. \tag{3.6.5}$$

If we assume the solution of the form

$$\kappa(n, m) = c_1 A^{(n)} z_1^m \tag{3.6.6}$$

and insert it into (3.6.5), we have

$$A^{(n)} = -\frac{c_1 z_1^n}{1 + e^{2\delta} z_1^{2(n+1)}} \tag{3.6.7}$$

where

$$\left.\begin{aligned} e^\delta &= \frac{c_1(t)}{\sqrt{1 - z_1^2}} = e^{\delta_0 + \beta t} \\ e^{\delta_0} &= \frac{c_1(0)}{\sqrt{1 - z_1^2}} \end{aligned}\right\} . \tag{3.6.8}$$

Then, by (3.4.22), we have

$$\begin{aligned} [K(n, n)]^{-2} &= 1 + c_1^2 z_1^{2n} + c_1^3 A^{(n)} z_1^n \sum_{n'=n+1}^{\infty} z_1^{2n'} \\ &= \frac{1 + e^{2\delta} z_1^{2n}}{1 + e^{2\delta} z_1^{2(n+1)}} = \frac{1 + e^{2\delta - 2n\gamma}}{1 + e^{2\delta - 2(n+1)\gamma}} \,. \end{aligned} \tag{3.6.9}$$

In passing we note that

$$\left.\begin{aligned} [K(n, n)]^{-2} &\to 1 & (n \to +\infty) \\ [K(n, n)]^{-2} &\to z_1^{-2} & (n \to -\infty) \end{aligned}\right\} . \tag{3.6.10}$$

Equations (3.2.23; 3.6.2, 8) give

$$e^{-(Q_n - Q_{n-1})} - 1 = \left[\frac{K(n, n)}{K(n-1, n-1)}\right]^2 - 1$$

$$= \frac{(e^\gamma - e^{-\gamma})^2}{(e^{\gamma n - \delta} + e^{-\gamma n + \delta})^2}$$

$$= \beta^2 \operatorname{sech}^2(\gamma n - \beta t - \delta_0) \tag{3.6.11}$$

which is a soliton solution of (2.4.5). It moves to the right when $\beta > 0$ ($z_1 > 0$), and to the left when $\beta < 0$ ($z_1 < 0$).

An N soliton solution is given when we have $R(z) = 0$ and a spectrum of N bound states z_j ($j = 1, 2, \ldots, N$). In this case the kernel of the GL equation is

$$F(m) = \sum_{j=1}^N c_j^2 z_j^m \tag{3.6.12}$$

and the GL equation reduces to

$$\kappa(n, m) + \sum_{j=1}^N c_j^2 z_j^{n+m} + \sum_{n'=1}^\infty \kappa(n, n') \sum_{j=1}^N c_j^2 z_j^{n'+m} = 0 \qquad (m > n) . \tag{3.6.13}$$

The solution to this equation can be obtained by putting

$$\kappa(n, m) = \sum_j c_j A_j^{(n)} z_j^m \tag{3.6.14}$$

and inserting it into the GL equation to obtain

$$\sum_j c_j z_j^m (A_j^{(n)} + c_j z_j^n) + \sum_{n'=n+1}^\infty \sum_{i=1}^N \sum_{j=1}^N c_j^2 z_j^{n'+m} c_i A_i^{(n)} z_i^{n'}$$

$$= \sum_j c_j z_j^m \left[A_j^{(n)} + c_j z_j^n + \sum_{i=1}^N c_i c_j A_i^{(n)} \frac{(z_i z_j)^{n+1}}{1 - z_i z_j}\right] = 0 . \tag{3.6.15}$$

Thus we have

$$\sum_{i=1}^N \left[\delta_{ij} + c_i c_j \frac{(z_i z_j)^{n+1}}{1 - z_i z_j}\right] A_i^{(n)} = -c_j z_j^n \qquad (j = 1, 2, \cdots, N) , \tag{3.6.16}$$

and solving the simultaneous equations for $A_j^{(n)}$ we have

$$A_i^{(n)} = \frac{\det B(n)^{(i)}}{\det B(n)} , \tag{3.6.17}$$

where $B(n)$ is a symmetric matrix with the elements

$$[B(n)]_{ij} = \delta_{ij} + c_i c_j \frac{(z_i z_j)^{n+1}}{1 - z_i z_j} \tag{3.6.18}$$

and $B(n)^{(i)}$ is the matrix obtained by replacing the ith column of $B(n)$ by $-(c_1 z_1^n, \ldots, c_N z_N^n)^T$.

Then (3.4.22) becomes

$$[K(n, n)]^{-2} = 1 + \sum_j c_j^2 z_j^{2n} + \sum_i \sum_j c_i c_j^2 A_i^{(n)} z_j^n \frac{(z_i z_j)^{n+1}}{1 - z_i z_j} . \tag{3.6.19}$$

However, if we multiply the jth equation of (3.6.16) by $c_j z_j^n$ and sum over j, we have

$$\sum_j c_j A_j^{(n)} z_j^n + \sum_j c_j^2 z_j^{2n} + \sum_i \sum_j c_i c_j^2 A_i^{(n)} z_j^n \frac{(z_i z_j)^{n+1}}{1 - z_i z_j} = 0 . \tag{3.6.20}$$

Therefore (3.6.19) is simplified as

$$[K(n, n)]^{-2} = 1 - \sum_{j=1}^{N} c_j A_j^{(n)} z_j^n . \tag{3.6.21}$$

Further, the right-hand side of this equation can be rewritten in the following way. First, we note that

$$\left.\begin{aligned} [B(n-1)]_{ij} &= \delta_{ij} + c_i c_j \frac{(z_i z_j)^n}{1 - z_i z_j} \\ [B(n)]_{ij} &= \delta_{ij} + c_i c_j \frac{(z_i z_j)^n}{1 - z_i z_j} z_i z_j \end{aligned}\right\} \tag{3.6.22}$$

and that

$$[B(n-1) - B(n)]_{ij} = c_i z_i^n \cdot c_j z_j^n . \tag{3.6.23}$$

Then, we have

$$\det B(n-1) = \det \{[B(n)]_{ij} + c_i z_i^n \cdot c_j z_j^n\}$$

$$= \begin{vmatrix} B_{11} + c_1 z_1^n \cdot c_1 z_1^n & B_{12} + c_1 z_1^n \cdot c_2 c_2^n & \cdots \\ B_{21} + c_2 z_2^n \cdot c_1 z_1^n & B_{22} + c_2 z_2^n \cdot c_2 z_2^n & \cdots \\ \cdots\cdots\cdots\cdots\cdots \end{vmatrix}$$

$$= \det B(n) + c_1 z_1^n \begin{vmatrix} c_1 z_1^n & B_{12} & B_{13} & \cdots \\ c_2 z_2^n & B_{22} & B_{23} & \cdots \\ \cdots\cdots\cdots\cdots \end{vmatrix}$$

$$+ c_2 c_2^n \begin{vmatrix} B_{11} & c_1 z_1^n & B_{13} & \cdots \\ B_{21} & c_2 z_2^n & B_{23} & \cdots \\ \cdots\cdots\cdots\cdots \end{vmatrix} + \cdots$$

$$= \det B(n) - \sum_j c_j z_j^n \det B(n)^{(j)}$$

$$= \det B(n) \left(1 - \sum_j c_j z_j^n A_j^{(n)}\right) \tag{3.6.24}$$

and so

$$[K(n, n)]^{-2} = \frac{\det B(n-1)}{\det B(n)}. \tag{3.6.25}$$

Therefore, by (3.2.23), we obtain

$$e^{-(Q_n - Q_{n-1})} = \left[\frac{K(n, n)}{K(n-1, n-1)]}\right]^2$$

$$= \frac{\det B(n) \det B(n-2)}{[\det B(n-1)]^2}. \tag{3.6.26}$$

Recalling (1.4.18) or

$$Q_{n+1} - Q_n = - S_{n+1} + 2S_n - S_{n-1}, \tag{3.6.27}$$

we get

$$S_n = \ln \det B(n) + \text{a term linear in } n. \tag{3.6.28}$$

Further, using (1.4.7) we obtain

$$P_n = \text{const.} + \frac{d}{dt} \ln \frac{\det B(n-1)}{\det B(n)}. \tag{3.6.29}$$

Here, in the elements (3.6.18) of $B(n)$, the time dependence of the coefficients is given as

$$\left. \begin{aligned} c_j &= c_j(0)\, e^{\beta_j t} \\ \beta_j &= \pm \sinh \gamma_j \end{aligned} \right\} \tag{3.6.30}$$

and

$$z_j = \pm e^{-\gamma_j}. \tag{3.6.31}$$

A similar calculation of determinants was done for the Schrödinger equation by *Kay* and *Moses* [3.5a].

Problem 3.1. Show that, when there is a soliton, we have

$$\phi(n, z) = K(n, n)\, z^n \left[1 + c_1 A^{(n)} \frac{z_1^{n+1}}{z^{-1} - z_1}\right]$$

so that $(z = e^{ik}, z_1 = \text{real}, |z_1| < 1)$

$$\phi(n, z^{-1}) \to e^{-ikn} \qquad (n \to +\infty)$$
$$\phi(n, z^{-1}) \to e^{-ikn-i\eta} \qquad (n \to -\infty) ,$$

where

$$e^{-i\eta} = \frac{1}{\alpha(z)} = \frac{1 - z_1 z}{z_1 - z} ,$$

and for $|z| \leq 1$

$$\phi(n, z) = K(n, n)\left[z^n + \sum_{m=n+1}^{\infty} c_1 A^{(n)} z_1^m z^m\right]$$
$$= \sum_{m=n}^{\infty} K(n, m) z^m$$

with [cf. (3.6.6)]

$$K(n, m) = K(n, n) \kappa(n, m)$$
$$= K(n, n) c_1 A^{(n)} z_1^m .$$

Further verify, using (3.6.9, 7), that

$$[K(n, n)]^{-2} = 1 - c_1 A^{(n)} z_1^n$$

and that, for $z \to 0$ [cf. (3.4.14)]

$$\phi(n, z^{-1}) \simeq K(n, n) z^{-n}(1 - c_1 A^{(n)} z_1^n)$$

$$= \frac{1}{K(n, n)} z^{-n} .$$

Problem 3.2. Show that, when there is a soliton, the bound state eigenfunction is

$$\zeta(n, z_1) = c_1 \phi(n, z_1)$$

with

$$\phi(n, z_1) = K(n, n) z_1^n\left[1 + c_1 A^{(n)} z_1^n \frac{z_1^2}{1 - z_1^2}\right],$$

so that

$$\zeta(n, z_1) = (\pm)^n c_1(0) e^{-\delta_0} \frac{e^\gamma}{2} K(n, n) \text{ sech } [\gamma(n + 1) - \beta t - \delta_0] \quad (z_1 = \pm e^{-\gamma})$$

and

$$\zeta(n, z_1) \rightarrow \begin{cases} c_1(0)\, z_1^n e^{\beta t} & (n \rightarrow +\infty) \\ c_1(0)\, e^{-2\delta_0} z_1^{-n-1} e^{-\beta t} & (n \rightarrow -\infty). \end{cases}$$

Remark 1. Compare (3.2.12–14) and the description that followed.

Remark 2. For $z_1 > 0$ (a soliton moving to the right) $\zeta(n, z)$ asymptotically decreases monotonously while for $z_1 < 0$ (a soliton moving to the left) it decreases oscillating. This asymmetry has no essential meaning, and can be removed by using a complex hermitian matrix L [3.6].

Problem 3.3. For the case of a soliton, (3.6.18) gives

$$B(n) = 1 + c_1^2 \frac{z^{2(n+1)}}{1 - z_1^2}.$$

By comparing (3.6.28) and (2.4.4), show that

$$B(n) = 1 + A e^{-2(\gamma n - \beta t)}$$

where A is a constant.

3.7 The Relationship Between the Conserved Quantities and the Transmission Coefficient

As a matter of course, the conserved quantities are related to many other quantities. But as it is cumbersome to give a detailed discussion on these relationships, we only refer to that of the transmission coefficient $[\alpha(z)]^{-1}$ as an example [3.4].

First, we rewrite (3.2.1) for $\phi_n = \phi(n)$ and $\psi_n = \psi(n)$ as

$$\left. \begin{array}{ll} \sum_{n'} L_{nn'}\phi_n = \lambda\phi_n, & \phi_n \rightarrow z^n \quad (n \rightarrow +\infty) \\ \sum_{n'} L_{nn'}\psi_{n'} = \lambda\psi_n, & \psi_n \rightarrow z^{-n} \quad (n \rightarrow -\infty) \end{array} \right\} \tag{3.7.1}$$

or

$$\left. \begin{array}{l} a_{n-1}\phi_{n-1} + a_n\phi_{n+1} + b_n\phi_n = \lambda\phi_n \\ a_{n-1}\psi_{n-1} + a_n\psi_{n+1} + b_n\psi_n = \lambda\psi_n \end{array} \right\}. \tag{3.7.2}$$

If the first of these equations is multiplied by $-\psi_n$ and the second by ϕ_n, then added together, we have

$$w \equiv a_{n-1}(\psi_{n-1}\phi_n - \psi_n\phi_{n-1})$$
$$= a_n(\psi_n\phi_{n+1} - \psi_{n+1}\phi_n) = \text{(independent of } n) \,. \tag{3.7.3}$$

w is the discrete Wronskian (multiplied by a_n) which is independent of n, and already appeared in (3.3.19). Thus we have the relation

$$w = \frac{\alpha(z)}{2}(z - z^{-1}) \,. \tag{3.7.4}$$

Now, we rewrite w using the first of equations (3.7.2) as

$$w = a_{n-1}\psi_{n-1}\phi_n + (a_n\phi_{n+1} + b_n\phi_n - \lambda\phi_n)\,\psi_n \,. \tag{3.7.5}$$

We have, on the other hand, by multiplying the first equation of (3.7.2) by ψ_m,

$$a_{n-1}\phi_{n-1}\psi_m + (a_n\phi_{n+1} + b_n\phi_n - \lambda\phi_n)\,\psi_m = 0 \tag{3.7.6}$$

and by multiplying the second by ϕ_m,

$$a_{n-1}\psi_{n-1}\phi_m + (a_n\psi_{n+1} + b_n\psi_n - \lambda\psi_n)\,\phi_m = 0 \,. \tag{3.7.7}$$

Therefore if we define

$$G_{n,m} = G(z, n, m) \begin{cases} \dfrac{\psi_n\phi_m}{w} & (n \leq m) \\[2mm] \dfrac{\phi_n\psi_m}{w} & (n \geq m) \,, \end{cases} \tag{3.7.8}$$

(3.7.5) becomes

$$a_{n-1}G_{n-1,n} + a_n G_{n+1,n} + b_n G_{n,n} - \lambda G_{n,n} = 1 \,, \tag{3.7.9}$$

while (3.7.6) for $n - 1 \geq m$ and (3.7.7) for $n + 1 \leq m$ can be unified into

$$a_{n-1}G_{n-1,m} + a_n G_{n+1,m} + b_n G_{n,m} - \lambda G_{n,m} = 0 \qquad (n \neq m) \,. \tag{3.7.10}$$

These are further united into

$$\sum_{n'} L_{nn'}G_{n',m} - \lambda G_{n,m} = \delta_{nm} \tag{3.7.11}$$

or in a matrix form

$$(L - \lambda I)\,G = I \tag{3.7.12}$$

where I is a unit matrix. Thus we see that G is Green's function of $L - \lambda I$.

Returning to (3.7.1), we express the nth component by a subscript n, to have

$$(L - \lambda I)\, \psi\, |_n = 0, \qquad \lambda = \frac{z + z^{-1}}{2}\, . \tag{3.7.13}$$

If we differentiate this equation with respect to z, we have ($\dot\psi_n = d\psi_n/dz$, etc.)

$$(L - \lambda I)\, \dot\psi\, |_n = a_{n-1}\dot\psi_{n-1} + a_n\dot\psi_{n+1} + b_n\dot\psi_n - \lambda\dot\psi_n = \frac{1 - z^{-2}}{2}\, \psi_n\, . \tag{3.7.14}$$

We combine this with the equation

$$(L - \lambda I)\, \phi\, |_n = a_{n-1}\phi_{n-1} + a_n\phi_{n+1} + b_n\phi_n - \lambda\phi_n = 0 \tag{3.7.15}$$

for ϕ, by adding (3.7.14) multiplied by ϕ_n and (3.7.15) multiplied by $\dot\psi_n$, to obtain

$$U_{n-1} - U_n = \frac{1 - z^{-2}}{2}\, \psi_n\phi_n \tag{3.7.16}$$

with

$$U_n = a_n(\phi_{n+1}\dot\psi_n - \phi_n\dot\psi_{n+1})\, . \tag{3.7.17}$$

Then, summing over n from $-N + 1$ to N, we have

$$\sum_{n=-N+1}^{N} \phi_n\psi_n = \frac{2}{z^{-2} - 1} (U_N - U_{-N})\, . \tag{3.7.18}$$

Using (3.3.1–4) we have the estimation for $N \gg 1$:

$$U_N = \frac{1}{2} \{ -\beta z^{2N} + [-N + (N + 1)\, z^{-2}]\, \alpha + (z - z^{-1})\, \dot\alpha\}$$

$$U_{-N} = \frac{1}{2} \{\bar\beta z^{2N-2} + [N - (N - 1)\, z^{-2}]\, \bar\alpha\} \tag{3.7.19}$$

when $\bar\alpha = \alpha$ by (3.3.6). Thus, for

$$|z| < 1 \tag{3.7.20}$$

and for $N \gg 1$, we have

$$\sum_{n=-N+1}^{N} \phi_n\psi_n = 2N\alpha - z\dot\alpha\, . \tag{3.7.21}$$

Now, we denote by L_0 the value of L when $a_n = 1/2$ and $b_n = 0$, and by G^0 the corresponding value of G. Then

$$G^0(z, n, m) = \begin{cases} \dfrac{z^{-n+m}}{w_0} & (n \le m) \\[3mm] \dfrac{z^{-m+n}}{w_0} & (n \ge m) \end{cases} \qquad\qquad (3.7.22)$$

with

$$w_0 = \frac{z - z^{-1}}{2}. \qquad\qquad (3.7.23)$$

We write the inverse of (3.7.12), the resolvent, as

$$\left.\begin{aligned} G &= \frac{I}{L - \lambda I} \equiv R_\lambda \\[3mm] G^0 &= \frac{I}{L_0 - \lambda I} \equiv R_\lambda^0 \end{aligned}\right\}. \qquad\qquad (3.7.24)$$

The sum of the diagonal elements yields

$$\begin{aligned} \text{tr}\,(R_\lambda - R_\lambda^0) &= \sum_{m=-\infty}^{\infty} [G(z, m, m) - G^0(z, m, m)] \\[2mm] &= \sum_{m=-\infty}^{\infty} \left[\frac{\psi(z, m)\,\phi(z, m)}{w} - \frac{1}{w_0}\right] \\[2mm] &= \sum_{m=-\infty}^{\infty} \left[\frac{2}{(z - z^{-1})\,\alpha(z)}\,\psi(z, m)\,\phi(z, m) - \frac{2}{z - z^{-1}}\right] \\[2mm] &= \frac{2z}{(z^2 - 1)\,\alpha(z)} \lim_{N\to\infty} \sum_{m=-N+1}^{N} [\psi(z, m)\,\phi(z, m) - \alpha(z)] \\[2mm] &= -\frac{2z^2}{z^2 - 1}\frac{\dot\alpha(z)}{\alpha(z)} \\[2mm] &= -\frac{d}{d\lambda}\log\alpha(z), \qquad\qquad (3.7.25) \end{aligned}$$

where we have used (3.7.21).

In order to integral (3.7.25), we expand the left-hand side in powers of λ^{-1}, and obtain

$$\begin{aligned} \text{tr}\,(R_\lambda - R_\lambda^0) &= \text{tr}\,\{(L - \lambda I)^{-1} - (L_0 - \lambda I)^{-1}\} \\[2mm] &= -\frac{1}{\lambda}\,\text{tr}\,\left\{\left(1 + \frac{L}{\lambda} + \frac{L^2}{\lambda^2} + \cdots\right) - \left(1 + \frac{L_0}{\lambda} + \frac{L_0^2}{\lambda^2} + \cdots\right)\right\} \\[2mm] &= -\text{tr}\,\left\{\frac{L - L_0}{\lambda^2} + \frac{L^2 - L_0^2}{\lambda^3} + \cdots\right\}. \qquad\qquad (3.7.26) \end{aligned}$$

Therefore integrating (3.7.25) over λ from $-\infty$ to λ, we have

$$- \sum_{p=1}^{\infty} \frac{1}{p\lambda^p} \, \text{tr} \, \{L^p - L_0^p\} = \ln \frac{\alpha(z)}{\alpha(0)} \, . \tag{3.7.27}$$

As we have seen from (3.5.8), $\alpha(z)$ is independent of time. Therefore the coefficients of λ^{-p} in the expansion of $\log\alpha(z)$ give conserved quantities $\text{tr} \, \{L^p - L_p^0\}$. The coefficients in the expansion of $\log\alpha(z)$ in powers of z are also conserved quantities. This can be shown directly as follows.

We make use of the Poisson-Jensen formula

$$\ln |\alpha(z)| = \frac{1}{2\pi} \int_0^{2\pi} \ln |\alpha(e^{i\varphi})| \, \text{Re} \left\{ \frac{e^{i\varphi} + z}{e^{i\varphi} - z} \right\} d\varphi + \sum_{j=0}^{N} \ln \left| \frac{z - z_j}{1 - z_j z} \right| \tag{3.7.28}$$

which holds for $|z| < 1$. Here z_j $(j = 1, 2, \ldots, N)$ are the zeros of $\alpha(z)$. For the range $-1 < z < 1$, we have real $\alpha(z)$, so that $\ln |\alpha(z)| = \ln\alpha(z)$. Further for $|z| = 1$, by (3.3.10), we have

$$|\alpha(z)|^{-2} = 1 - |R(z)|^2 \, . \tag{3.7.29}$$

Therefore, if we rewrite the left-hand side of (3.7.28) as

$$\ln |\alpha(z)| = \ln |\alpha(0)| + \sum_{p=1}^{\infty} K_p z^p \, , \tag{3.7.30}$$

K_p's are conserved quantities, which are given by expanding the right-hand side of (3.7.28) as

$$K_p = - \frac{1}{2\pi} \int_0^{2\pi} \ln (1 - |R(e^{i\varphi})|^2) \cos (p\varphi) \, d\varphi + \sum_{j=0}^{N} \frac{1}{p} (z_j^p - z_j^{-p})$$

$$(p \neq 0) \, . \tag{3.7.31}$$

In the right-hand side of the above equation, the first term is related to the continuum spectrum and the second term to the discrete spectrum.

Problem 3.4. By using $\lambda = (z + z^{-1})/2$, from (3.7.27) we have

$$\ln \frac{\alpha(z)}{\alpha(0)} = - 2 \, \text{tr} \, \{L - L_0\} \cdot z + 2 \, \text{tr} \, \{L^2 - L_0^2\} \cdot z^2 + \cdots \, .$$

Verify that the conserved quantities given by the expansion coefficients are

$$- 2 \, \text{tr} \, \{L - L_0\} = - 2 \sum b_n = \sum P_n = P$$
$$2 \, \text{tr} \, \{L^2 - L_0^2\} = \sum [(4a_n^2 - 1) + 2b_n^2] = E$$

where P stands for the total momentum, and E the total energy,

$$E = \frac{1}{2} \sum_n P_n^2 + \sum_n (e^{-(Q_{n+1} - Q_n)} - 1) \, .$$

Hint: $\lambda^{-1} = \dfrac{2z}{1 + z^2} = 2z(1 - z^2 + \cdots)$.

Problem 3.5. For a soliton (cf. Sect. 3.6 [Problem 3.1])

$$\alpha(z) = \frac{z_1 - z}{1 - zz_1} \ .$$

Using this and the results of the preceding problem, show that

$$P = -(z_1 - z_1^{-1}) = 2\sinh \kappa \qquad (z_1 = e^{-\kappa})$$

$$E = \frac{1}{2}(z_1^{-2} - z_1^2) = \sinh 2\kappa \ .$$

These are respectively equivalent to $-K_1$ and K_2.

Remark: The energy of a soliton given in Sect. 2.4 **Problem 2.9** includes the contribution $(-2a/b)\kappa$ due to the attractive term ar of the interaction potential of the lattice.

3.8 Extensions of the Equations of Motion and the Kac-Moerbeke System

We have seen that if the matrix L of the Lax formalism has only the diagonal and the first off-diagonal elements, while B has only the first off-diagonal elements, it gives the lattice with exponential interaction. This choice of L and B will be the simplest. We may leave L and take the more complicated B. In this case, $BL - LB$ must have only the diagonal and the first off-diagonal elements like L or dL/dt. For simplicity, we assume vanishing elements on the upper right and the lower left corners for a while, and do not ask its mechanical implication (cf. Sect. 3.10). Let L be

$$L = \begin{pmatrix} b_1 & a_1 & & & \\ a_1 & b_2 & a_2 & & 0 \\ & a_2 & b_3 & \ddots & \\ & 0 & & \ddots & \end{pmatrix} \tag{3.8.1}$$

and assume

$$B_2 = \begin{pmatrix} 0 & \beta_1 & \gamma_1 & & \\ -\beta_1 & 0 & \beta_2 & \gamma_2 & 0 \\ -\gamma_1 & -\beta_2 & 0 & \ddots & \\ & 0 & & \ddots & \end{pmatrix} . \tag{3.8.2}$$

Then, by the requirement that the $(k - 1, k + 2)$ element of $B_2 L - LB_k$ should vanish, we have $a_{k-1} \gamma_k = \gamma_{k-1} a_{k+1}$. If we assume that γ_k's do not vanish, this is satisfied by setting

$$\gamma_k = a_k a_{k+1} . \tag{3.8.3}$$

In a similar way as the $(k - 1, k + 1)$ element should vanish, we have $a_k \beta_{k-1} + \gamma_{k-1} b_{k+1} = \gamma_{k-1} b_{k-1} + a_{k-1} \beta_k$ and this is then satisfied by setting [3.7]

$$\beta_k = (b_k + b_{k+1}) a_k . \tag{3.8.3'}$$

In this case the motion is also isospectral and integrable, though we cannot have any mechanical interpretation of the equations of motion. *Moser* [3.7] proved the existence of B_p which has up to the pth off-diagonal element.

For a simpler case, we put $b_n = 0$ to have

$$\mathscr{L} = \begin{pmatrix} 0 & \alpha_1 & & & & \\ \alpha_1 & 0 & \alpha_2 & & 0 & \\ & \alpha_2 & 0 & & & \\ & & & \ddots & & \\ & & & & 0 & \alpha_{n-1} \\ & 0 & & & \alpha_{n-1} & 0 \end{pmatrix} \tag{3.8.4}$$

$$\mathscr{B} = \begin{pmatrix} 0 & 0 & \alpha_1 \alpha_2 & & & & \\ 0 & 0 & 0 & \alpha_2 \alpha_3 & & 0 & \\ -\alpha_1 \alpha_2 & 0 & 0 & & & & \\ & & & \ddots & & & \\ & & & 0 & 0 & \alpha_{n-2} \alpha_{n-1} \\ & 0 & & 0 & 0 & 0 \\ & & -\alpha_{n-2} \alpha_{n-1} & 0 & 0 \end{pmatrix} . \tag{3.8.5}$$

Then the equation of motion

$$\frac{d\mathscr{L}}{dt} = \mathscr{B} \mathscr{L} - \mathscr{L} \mathscr{B} \tag{3.8.6}$$

gives

$$\dot{\alpha}_k = \alpha_k (\alpha_{k+1}^2 - \alpha_{k-1}^2) . \tag{3.8.7}$$

If we write

$$\alpha_k = \frac{1}{2} e^{u_k/2} , \tag{3.8.8}$$

we have

$$\dot{u}_k = \frac{1}{2} \left(e^{u_{k+1}} - e^{u_{k-1}} \right) . \tag{3.8.9}$$

If we further write

$$t/2 \rightarrow t, \qquad u_k = - R_k , \tag{3.8.10}$$

we have

$$\dot{R}_k = e^{-R_{k+1}} - e^{-R_{k-1}} . \tag{3.8.11}$$

This sort of equation was studied by *Kac* and *van Moerbeke* using the inverse scattering method [3.8]; for convenience we shall refer to it as the KM system. Though it admits no direct mechanical interpretation, KM noted that it has a simple relation to the lattice with exponential interaction.

To show the relation between the KM system and the lattice, we put

$$\left.\begin{aligned} R_{2k} &= x_{2k+1} - x_{2k} \\ R_{2k+1} &= x_{2k+2} - x_{2k+1} \end{aligned}\right\} . \tag{3.8.12}$$

Figure 3.1 illustrates the relation between $x_{2k}, \ldots$ and $R_{2k}, \ldots$ (3.8.11) is satisfied by

$$\left.\begin{aligned} \dot{x}_{2k} &= e^{-R_{2k-1}} + e^{-R_{2k}} - \alpha \\ \dot{x}_{2k+1} &= e^{-R_{2k}} + e^{-R_{2k+1}} - \alpha \end{aligned}\right\} \tag{3.8.13}$$

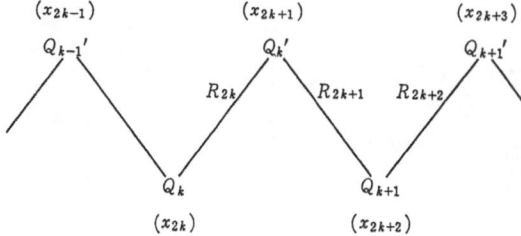

(x_{2k-1}) (x_{2k+1}) (x_{2k+3})

Q_{k-1}' Q_k' Q_{k+1}'

R_{2k} R_{2k+1} R_{2k+2}

Q_k Q_{k+1}

(x_{2k}) (x_{2k+2})

Fig. 3.1. Relation between a Kac-Moerbeke system and the Bäcklund transformation (cf. Sect. 3.9)

where α is a constant. Further we put (cf. Fig. 3.1)

$$x_{2k} = Q_k, \qquad x_{2k+1} = Q_k' \tag{3.8.14}$$

and thus

$$R_{2k} = Q_k' - Q_k, \qquad R_{2k+1} = Q_{k+1} - Q_k' \tag{3.8.15a}$$
$$R_{2k} + R_{2k+1} = Q_{k+1} - Q_k . \tag{3.8.15b}$$

Then (3.8.13) is rewritten as

$$
\left.\begin{aligned}
\dot{Q}_k &= e^{-R_{2k-1}} + e^{-R_{2k}} - \alpha \\
\dot{Q}_k' &= e^{-R_{2k}} + e^{-R_{2k+1}} - \alpha
\end{aligned}\right\}. \tag{3.8.16}
$$

Differentiating (3.8.16) with respect to t, we have

$$
\begin{aligned}
\ddot{Q}_k &= -\dot{R}_{2k-1}\, e^{-R_{2k-1}} - \dot{R}_{2k}\, e^{-R_{2k}} \\
&= (e^{-R_{2k-2}} - e^{-R_{2k}})\, e^{-R_{2k-1}} + (e^{-R_{2k-1}} - e^{-R_{2k+1}})\, e^{-R_{2k}} \\
&= e^{(R_{2k-2}+R_{2k-1})} - e^{-(R_{2k}+R_{2k+1})}
\end{aligned} \tag{3.8.17}
$$

and that, by (3.8.15),

$$
\ddot{Q}_k = e^{-(Q_k - Q_{k-1})} - e^{-(Q_{k+1} - Q_k)} . \tag{3.8.18a}
$$

This describes the lattice with exponential interaction. In the same way, we have

$$
\ddot{Q}_k' = e^{-(Q_k' - Q_{k-1}')} - e^{-(Q_{k+1}' - Q_k')} \tag{3.8.18b}
$$

which describes another lattice with the same exponential interaction. Thus, from the KM system, we can derive two lattices.

We may also consider the following. As (3.8.4, 5) imply that

$$
\left.\begin{aligned}
\mathscr{L}_{n,n+1} &= \mathscr{L}_{n+1,n} = \alpha_n \\
\mathscr{B}_{n,n+2} &= -\mathscr{B}_{n+2,n} = \alpha_n \alpha_{n+1} \\
\text{other elements} &= 0
\end{aligned}\right\}, \tag{3.8.19}
$$

we have

$$
\left.\begin{aligned}
(\mathscr{L}^2)_{n,n} &= \alpha_{n-1}^2 + \alpha_n^2 = \frac{1}{4}(e^{-R_{n-1}} + e^{-R_n}) \\
(\mathscr{L}^2)_{n,n+2} &= (\mathscr{L}^2)_{n+2,n} = \alpha_n \alpha_{n+1} = \frac{1}{4} e^{-(R_n + R_{n+1})/2}
\end{aligned}\right\} \tag{3.8.20}
$$

and, by using (3.8.14–16),

$$
\left.\begin{aligned}
(\mathscr{L}^2)_{2n,2n} &= \frac{1}{4}\dot{Q}_n + \frac{\alpha}{4} \\
(\mathscr{L}^2)_{2n+1,2n+1} &= \frac{1}{4}\dot{Q}_n' + \frac{\alpha}{4} \\
(\mathscr{L}^2)_{2n,2n+2} &= \frac{1}{4} e^{-(Q_{n+1} - Q_n)/2} \\
(\mathscr{L}^2)_{2n+1,2n+3} &= \frac{1}{4} e^{-(Q_{n+1}' - Q_n')/2}
\end{aligned}\right\}. \tag{3.8.21}
$$

Therefore, the equation of motion which holds for $\mathscr{L}^2$, that is,

$$\frac{d\mathscr{L}^2}{dt} = \mathscr{B}\mathscr{L}^2 - \mathscr{L}^2\mathscr{B} \tag{3.8.22}$$

includes the equations of motion for both Q_n (3.8.18a) and Q'_n (3.8.18b), or equations $dL/dt = BL - LB$ (3.1.7) and $dL'/dt = B'L' - L'B'$ where Q_n and P_n are replaced by Q'_n and P'_n.

If we put $\alpha = 0$, we have

$$\left.\begin{array}{ll} a_n = 2\alpha_{2n}\alpha_{2n+1}, & a'_n = 2\alpha_{2n+1}\alpha_{2n+2} \\ b_n = 2(\alpha_{2n-1}^2 + \alpha_{2n}^2), & b'_n = 2(\alpha_{2n}^2 + \alpha_{2n+1}^2) \end{array}\right\}. \tag{3.8.23}$$

Problem 3.6. Verify γ_k and β_k of (3.8.3). Show that if we assume that $\gamma_k = 0$, we have $\beta_k \propto a_k$.

Problem 3.7. Verify a particular solution to (3.8.11) of the form

$$R_{2n} = \ln\left|\frac{\cosh(\kappa n + \beta t)}{(\cosh[\kappa(n-1) + \beta t]}\right|$$

$$R_{2n+1} = \ln\left|\frac{\cosh(\kappa n + \beta t)}{\cosh[\kappa(n+1) + \beta t]}\right|,$$

or

$$e^{-R_{2n}} = \frac{1}{\sqrt{1-\mu^2}}[1 - \mu\tanh(\kappa n + \beta t)]$$

$$e^{-R_{2n+1}} = \frac{1}{\sqrt{1-\mu^2}}[1 + \mu\tanh(\kappa n + \beta t)]$$

with

$$\cosh\kappa = 1/\sqrt{1-\mu^2}, \qquad \beta = \sinh\kappa.$$

Further show that these lead to a soliton for the lattice with exponential interaction, cf. (2.4.2).

Problem 3.8. Show that (3.8.16) has a particular solution

$$Q_k = \gamma = \text{const.},$$

$$e^{Q_n'-\gamma} = \frac{\cosh(\kappa n + \beta t)}{\cosh[\kappa(n+1) + \beta t]},$$

with

$$\beta = \sinh\kappa, \qquad \alpha = e^\kappa + e^{-\kappa}.$$

3.9 The Bäcklund Transformation

In the preceding section we have seen that two lattices are joined by the KM system. This means that if a motion in a lattice is transformed by the KM system we obtain another motion in the same lattice. In general, a transformation which transforms a solution to another solution is called the Bäcklund transformation. In the present case, the transformation induced by the KM system is a Bäcklund transformation of the lattice with exponential interaction. We can show that this is a canonical transformation in the following way.

I) We start with a simple comment. For a set of variables $Q = \{Q_n\}$ and $P = \{P_n\}$ we consider a transformation

$$\left.\begin{array}{l} P_n = A \dfrac{f_n(Q)}{f_n(Q')} + \dfrac{f_{n-1}(Q')}{Af_n(Q)} \\[3mm] P'_n = A \dfrac{f_n(Q)}{f_n(Q')} + \dfrac{f_n(Q')}{Af_{n+1}(Q)} \end{array}\right\} \tag{3.9.1}$$

where A is a constant. Then we see

$$\left[\frac{1}{2}\sum_{n=-N_0}^{N} P'^2_n + \sum_{n=-N_0}^{N} \frac{f_{n-1}(Q')}{f_n(Q')}\right] - \left[\frac{1}{2}\sum_{n=-N_0}^{N} P^2_n + \sum_{n=-N_0}^{N} \frac{f_n(Q)}{f_{n+1}(Q)}\right]$$

$$= -\frac{1}{2A^2}\left[\frac{f_{-N_0-1}(Q')}{f_{-N_0}(Q)}\right]^2 + \frac{1}{2A^2}\left[\frac{f_N(Q')}{f_{N+1}(Q)}\right]^2 . \tag{3.9.2}$$

Thus, if we have a canonical transformation which gives (3.9.1), it transforms the Hamiltonian

$$H(Q, P) = \frac{1}{2}\sum_n P^2_n + \sum_n \frac{f_n(Q)}{f_{n+1}(Q)} \tag{3.9.3}$$

to itself, provided that the boundary conditions ensure the constancy of the right-hand side of (3.9.2).

The Hamiltonian for the lattice with exponential interaction

$$H(Q, P) = \frac{1}{2}\sum_{n=-\infty}^{\infty} P^2_n + \sum_{n=-\infty}^{\infty} e^{-(Q_{n+1}-Q_n)} , \tag{3.9.4}$$

has this form if $f_n(Q) = e^{Q_n}$. Therefore, we take the generating function

$$W(Q, Q') = \sum_j \left[Ae^{-(Q_{j'}-Q_j)} - \frac{1}{A}e^{-(Q_{j+1}-Q_{j'})} + \alpha(Q'_j - Q_j)\right] \tag{3.9.5}$$

for the canonical transformation [A, α are constants, for positive (negative) A, we may take $A = 1$ ($A = -1$), by shifting all the Q_n's by an appropriate constant]. Then the canonical transformation is given as

$$P_n = \frac{\partial W}{\partial Q_n} = Ae^{-(Q_n'-Q_n)} + \frac{1}{A}\,e^{-(Q_n-Q_{n-1}')} - \alpha \left.\begin{array}{c}\\\\\end{array}\right\}.$$
$$P_n' = -\frac{\partial W}{\partial Q_n'} = Ae^{-(Q_n'-Q_n)} + \frac{1}{A}\,e^{-(Q_{n+1}-Q_n')} + \alpha$$

$$(3.9.6)$$

If we omit the additional constant α, this has the same form as (3.9.1). As the boundary conditions we assume that Q_n and Q_n' take constant values at infinity. Namely

$$\begin{array}{llc} Q_n \to Q_{-\infty}, & Q_n' \to Q_{-\infty}' & (n \to -\infty) \\ Q_n \to Q_\infty, & Q_n' \to Q_\infty' & (n \to +\infty) \end{array}\left.\right\}.$$

$$(3.9.7)$$

Since, then, the nomenta P_n and P_n' vanish at infinity, we have

$$Ae^{(-Q_{-\infty}'-\infty)} + \frac{1}{A}\,e^{-(Q_{-\infty}-Q_{-\infty}')} - \alpha = 0 \left.\begin{array}{c}\\\\\end{array}\right\}$$
$$Ae^{-(Q_\infty'-Q_\infty)} + \frac{1}{A}\,e^{-(Q_\infty-Q_\infty')} - \alpha = 0$$

$$(3.9.8)$$

which give relations between α, A, $Q_{-\infty}$, etc. That is,

$$A = \exp\left(\frac{Q_\infty' + Q_{-\infty}'}{2} - \frac{Q_\infty + Q_{-\infty}}{2}\right) \left.\begin{array}{c}\\\\\\\\\end{array}\right\}$$
$$\alpha = \exp\left(\frac{Q_\infty' - Q_{-\infty}'}{2} - \frac{Q_\infty - Q_{-\infty}}{2}\right) + \exp\left(-\frac{Q_\infty' - Q_{-\infty}'}{2} + \frac{Q_\infty - Q_{-\infty}}{2}\right)$$

$$(3.9.9)$$

where we have assumed $A > 0$ (the case $A < 0$ can be treated in a similar way). From (3.9.6), we obtain

$$\sum_{n=-\infty}^{\infty} (P_n' - P_n) = 2\sinh\{[(Q_\infty' - Q_{-\infty}') - (Q_\infty - Q_{-\infty})]/2\}.$$

$$(3.9.10)$$

Therefore, we have

$$\sum_{n=-\infty}^{\infty} P_n' = \sum_{n=-\infty}^{\infty} P_n + \text{const}.$$

$$(3.9.11)$$

and, by (3.9.6),

$$\sum_{n=-\infty}^{\infty} [(P_n' + \alpha)^2 - (P_n + \alpha)^2] = 2\sum_{n=-\infty}^{\infty} (e^{-(Q_{n+1}-Q_n)} - e^{-(Q_n'-Q_{n-1}')}) + \text{const}.$$

$$(3.9.12)$$

Thus in view of (3.9.11), we see that

$$\frac{1}{2} \sum_{n=-\infty}^{\infty} P_n'^2 + \sum_{n=-\infty}^{\infty} e^{-(Q_n'-Q_{n-1}')} = \frac{1}{2} \sum_{n=-\infty}^{\infty} P_n^2 + \sum_{n=-\infty}^{\infty} e^{-(Q_{n+1}-Q_n)} + \text{const.} \, .$$

$$(3.9.13)$$

In general, a canonical transformation derived by a generating function W transforms the Hamiltonian to

$$H'(Q'P') = H[Q(Q', P'), P(Q', P')] + \frac{\partial W}{\partial t} \, . \tag{3.9.14}$$

Therefore the canonical transformation (3.9.5, 6) transforms the Hamiltonian (3.9.4) to a new Hamiltonian of the same form

$$H'(Q', P') = \frac{1}{2} \sum_{n=-\infty}^{\infty} P_n'^2 + \sum_{n=-\infty}^{\infty} e^{-(Q_n'-Q_{n-1}')} + \text{const.} \, . \tag{3.9.15}$$

In other words, this canonical transformation transforms a lattice with exponential interaction to the same lattice. That is, if a set (Q, P) gives a possible motion in the lattice, then the canonical transformation yields another possible motion (Q', P'). Since the equations of motion give us $\dot{Q}_n = \partial H/\partial P_n = P_n$ and $\dot{Q}_n' = \partial H'/\partial P_n' = P_n'$, (3.9.6) is equivalent to

$$\left. \begin{array}{l} \dot{Q}_n = A e^{-(Q'-Q_n)} + \dfrac{1}{A} e^{-(Q_n-Q_{n-1}')} - \alpha \\[3mm] \dot{Q}_n' = A e^{-(Q_n'-Q_n)} + \dfrac{1}{A} e^{-(Q_{n+1}-Q_n')} - \alpha \end{array} \right\} . \tag{3.9.16}$$

If a solution $Q_n = Q_n(t)$ is given, by solving (3.9.16) (which are differential equations of the first rank with respect to time), we can obtain another solution $Q_n' = Q_n'(t)$. The Bäcklund transformation connects different solutions by relations of lower rank. The first Bäcklund transformation was introduced in a study of the geometrical nature of the equation $\phi_{tt} - \phi_{xx} = \sin \phi$ (the sine-Gordon equation). Sometimes, relations between solutions of different equations are also referred to as the Bäcklund transformation. Equation (3.9.16) is a Bäcklund transformation for the lattice with exponential interaction, and can be considered as a canonical transformation. Further, comparing (3.9.16) with (3.8.15) we see that (3.8.16) is of the same form as the KM system, and that Fig. 3.1 shows the Bäcklund transformation for the lattice with exponential interaction.

The equations of motion for the lattice has a trivial solution

$$Q_n = \gamma = \text{const.} \, . \tag{3.9.17}$$

As we see from **Problem 3.8** of Sect. 3.8, (3.9.16) then gives

$$e^{Q_n'-\gamma'} = z \frac{\cosh [\kappa(n-1) + \beta t + \delta]}{\cosh [\kappa n + \beta t + \delta]} \tag{3.9.18a}$$

where δ is a phase constant, and

$$\left.\begin{array}{l} \beta = \sinh\kappa, \qquad \alpha = z + z^{-1} \\ z = Ae^{\gamma - \gamma'} = e^{-\kappa} \\ Q'_n \to Q'_{-\infty} = \gamma' \qquad (n \to -\infty) \end{array}\right\}, \tag{3.9.18b}$$

and we are given a soliton solution

$$e^{-(Q_{n+1}' - Q_n')} - 1 = \sinh^2\kappa \ \text{sech}^2(\kappa n + \beta t + \delta) . \tag{3.9.19}$$

If we change the phase constant δ by $\delta + \pi/2\mathrm{i}$, we have another solution

$$e^{-(Q_{n+1}' - Q_n')} - 1 = - \sinh^2\kappa \ \text{cosech}^2(\kappa n + \beta t + \delta) \tag{3.9.20}$$

which, though diverges (antisoliton), gives a partially possible motion of the lattice.

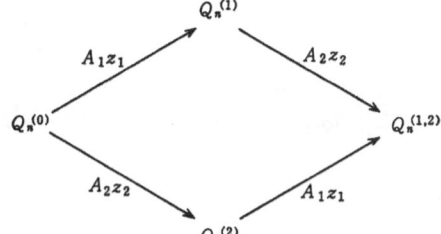

Fig. 3.2. Superposition of two Bäcklund transformations

II) We consider two successive Bäcklund transformations [3.10], characterized respectively by (A_1, z_1) and (A_2, z_2). We distinguish two routes starting from $Q_n^{(0)}$; one arrives at $Q_n^{(1)}$ by (A_1, z_1) and then at $Q_n^{(1,2)}$ by (A_2, z_2), and the other arrives at $Q_n^{(2)}$ by (A_2, z_2) and then at $Q_n^{(2,1)}$, and we demand that $Q_n^{(1,2)} = Q_n^{(2,1)}$ (cf. Fig. 3.2). We impose the boundary conditions

$$Q_{-\infty}^{(0)} = \gamma^{(0)}, \quad Q_{-\infty}^{(1)} = \gamma^{(1)}, \quad Q_{-\infty}^{(2)} = \gamma^{(2)}, \quad Q_{-\infty}^{(1,2)} = \gamma^{(1,2)} \tag{3.9.21}$$

for $n \to -\infty$ ($\gamma^{(0)}, \gamma^{(1)}, \gamma^{(2)}, \gamma^{(1,2)}$ are all constants). Writing down the transformations according to (3.9.16) we have

$$\left.\begin{array}{l} \dfrac{d}{dt}(Q_n^{(0)} - Q_{n-1}^{(1)}) = z_1 \left[e^{-(Q_n^{(1)} - \gamma^{(1)} - Q_n^{(0)} + \gamma^{(0)})} - e^{-(Q_{n-1}^{(1)} - \gamma^{(1)} - Q_{n-1}^{(0)} + \gamma^{(0)})} \right] \\[3mm] \dfrac{d}{dt}(Q_n^{(0)} - Q_{n-1}^{(2)}) = z_2 \left[e^{-(Q_n^{(2)} - \gamma^{(2)} - Q_n^{(0)} + \gamma^{(0)})} - e^{-(Q_{n-1}^{(2)} - \gamma^{(2)} - Q_{n-1}^{(0)} + \gamma^{(0)})} \right] \\[3mm] \dfrac{d}{dt}(Q_n^{(1)} - Q_{n-1}^{(1,2)}) = z_2 \left[e^{-(Q_n^{(1,2)} - \gamma^{(1,2)} - Q_n^{(1)} + \gamma^{(1)})} - e^{-(Q_{n-1}^{(1,2)} - \gamma^{(1,2)} - Q_{n-1}^{(1)} + \gamma^{(1)})} \right] \\[3mm] \dfrac{d}{dt}(Q_n^{(2)} - Q_{n-1}^{(1,2)}) = z_1 \left[e^{-(Q_n^{(1,2)} - \gamma^{(1,2)} - Q_n^{(2)} + \gamma^{(2)})} - e^{-(Q_{n-1}^{(1,2)} - \gamma^{(1,2)} - Q_{n-1}^{(2)} + \gamma^{(2)})} \right] \end{array}\right\}$$

$$\tag{3.9.22a}$$

with

$$
\left.
\begin{aligned}
z_1 &= A_1 e^{\gamma^{(0)}-\gamma^{(1)}} = A_1 e^{\gamma^{(2)}-\gamma^{(1,2)}} \\
z_2 &= A_2 e^{\gamma^{(0)}-\gamma^{(2)}} = A_2 e^{\gamma^{(1)}-\gamma^{(1,2)}} \\
\gamma^{(0)} &+ \gamma^{(1,2)} = \gamma^{(1)} + \gamma^{(2)}
\end{aligned}
\right\}.
$$

(3.9.22b)

If we eliminate time derivatives from (3.9.22a) we obtain

$$
\begin{aligned}
\Phi_n &\equiv [z_1 e^{-(Q_{n+1}(1)-\gamma^{(1)})} - z_2 e^{-(Q_{n+1}(2)-\gamma^{(2)})}] \, e^{+(Q_{n+1}(0)-\gamma^{(0)})} \\
&\quad + [z_2 e^{+(Q_n(1)-\gamma^{(1)})} - z_1 e^{+(Q_n(2)-\gamma^{(2)})}] \, e^{-(Q_n(1,2)-\gamma^{(1,2)})} \\
&= \Phi_{n-1} \, .
\end{aligned}
$$

(3.9.23)

Thus Φ_n is independent of n, and as $\Phi_n \to 0$ at $n \to -\infty$, Φ_n vanishes identically,

$$
\Phi_n = 0 \, .
$$

(3.9.24)

Therefore

$$
e^{Q_n(1,2)-\gamma(1,2)+Q_{n+1}(0)-\gamma(0)} = \frac{z_1 e^{Q_n(2)-\gamma(2)} - z_2 e^{Q_n(1)-\gamma(1)}}{z_1 e^{-(Q_{n+1}(1)-\gamma(1))} - z_2 e^{-(Q_{n+1}(2)-\gamma(2))}} \, ,
$$

(3.9.25)

or

$$
\begin{aligned}
&e^{Q_n(1,2)-\gamma(1,2)+Q_{n+1}(0)-\gamma(0)} \\
&= \frac{z_1 e^{Q_n(2)-\gamma(2)} - z_2 e^{Q_n(1)-\gamma(1)}}{z_1 e^{Q_{n+1}(2)-\gamma(2)} - z_2 e^{Q_{n+1}(1)-\gamma(1)}} \, e^{Q_{n+1}(1)-\gamma(1)+Q_{n+1}(2)-\gamma(2)}
\end{aligned}
$$

(3.9.26)

which means that if we are given $Q_n^{(0)}$, $Q_n^{(1)}$ and $Q_n^{(2)}$, we can derive a new solution $Q_n^{(1,2)}$ by an algebraic calculation. We may start from $Q_n^{(1)}$ or $Q_n^{(2)}$ or even $Q_n^{(1,2)}$, instead of from $Q_n^{(0)}$, and obtain other solutions algebraically by constructing a ladder of transformations. However, real calculation will become more and more elaborate as we proceed.

For example, we start from

$$
Q_n^{(0)} = \gamma^{(0)} = \text{const.}
$$

(3.9.27)

and show the procedure of obtaining a two-soliton solution $Q_n^{(1,2)}$. Extending (3.9.18a), we introduce $Q_n^{(1)}$ and $Q_n^{(2)}$ by

$$
\left.
\begin{aligned}
e^{Q_n(1)-\gamma(1)} &= z_1 \frac{\psi_1(n)}{\psi_1(n+1)} \\
e^{Q_n(2)-\gamma(2)} &= z_2 \frac{\psi_2(n)}{\psi_2(n+1)}
\end{aligned}
\right\}.
$$

(3.9.28)

Then, by (3.9.26), we obtain

$$e^{\varrho_n(1,2)-\gamma(1,2)} = z_1 z_2 \frac{\Psi_n}{\Psi_{n+1}}$$

(3.9.29)

with

$$\Psi_n = \psi_2(n)\psi_1(n+1) - \psi_1(n)\psi_2(n+1)$$

(3.9.30)

or, by rewriting,

$$e^{-(\varrho_{n+1}(1,2)-\varrho_n(1,2))} = \frac{\Psi_{n+2}\Psi_n}{\Psi_{n+1}^2}.$$

(3.9.31)

Therefore, we have

$$\left.\begin{array}{l} S_n = \ln \Psi_{n+1} + (\text{a term linear in } n) \\ e^{-(\varrho_{n+1}^{(1,2)}-\varrho_n^{(1,2)})} - 1 = \dfrac{d^2}{dt^2} S_n \end{array}\right\}.$$

(3.9.32)

For ψ_1, we insert from (3.9.18a)

$$\begin{aligned} \psi_1(n) &= 2D_1 \cosh\left[(n-1)\kappa_1 + \beta_1 t + \delta_1\right] \\ &= B_1 e^{-n\kappa_1 - \beta_1 t} + C_1 e^{n\kappa_1 + \beta_1 t}, \end{aligned}$$

(3.9.33a)

where

$$D_1^2 = B_1 C_1, \qquad e^{\delta_1} = \sqrt{\frac{C_1}{B_1}}\, e^{\kappa_1}, \qquad \beta_1 = \sinh \kappa_1$$

(3.9.33b)

and

$$\begin{aligned} \psi_2(n) &= 2D_2 \cosh\left[(n-1)\kappa_2 + \beta_2 t + \delta_2\right] \\ &= B_2 e^{-n\kappa_2 - \beta_2 t} + C_2 e^{n\kappa_2 + \beta_2 t}, \end{aligned}$$

(3.9.34a)

where

$$D_2^2 = B_2 C_2, \qquad e^{\delta_2} = \sqrt{\frac{C_2}{B_2}}\, e^{\kappa_2}, \qquad \beta_2 = \sinh \kappa_2.$$

(3.9.34b)

Then we obtain the two-soliton solution in terms of

$$\begin{aligned} \Psi_n &= 2D_+ \cosh\left[(n-1)(\kappa_1 + \kappa_2) + (\beta_1 + \beta_2)t + \delta_+\right] \\ &\quad + 2D_- \cosh\left[(n-1)(\kappa_2 - \kappa_1) + (\beta_2 - \beta_1)t + \delta_-\right] \end{aligned}$$

(3.9.35a)

with

$$2D_+^2 = -4D_1 D_2 \sinh^2 (\kappa_2 - \kappa_1)$$
$$2D_-^2 = -4D_1 D_2 \sinh^2 (\kappa_1 + \kappa_2)$$
$$e^{2\delta_+} = -\frac{C_1 C_2}{B_1 B_2} e^{3(\kappa_1 + \kappa_2)}$$
$$e^{2\delta_-} = -\frac{C_1 C_2}{B_1 B_2} e^{3(\kappa_2 - \kappa_1)}$$

(3.9.35b)

(3.9.35a) is of the form of (2.5.1) for a two-soliton solution. In order that $Q_n^{(1, 2)}$ includes no antisoliton, D_+ and D_- must have the same sign (real and both positive), which means that D_1 and D_2 have opposite signs. Thus we have a proper two-soliton solution when one of $Q_n^{(1)}$ and $Q_n^{(2)}$ is a soliton solution and the other is an antisoliton solution. We meet a similar situation when we apply the Bäcklund transformation to the KdV equation, and this will be a general property of the transformation. It turns out to be very hard if we try to see a concrete expression of a multisoliton solution.

Formally we may take the following alternative way [3.10]. Generalizing (3.9.18a), we put

$$e^{Q'_{n-1}} = e^{(Q_n + Q_{n-1})/2} \frac{\varphi(n-1)}{\varphi(n)}$$

(3.9.36)

and insert it into (3.9.16) to obtain

$$a_{n-1}\varphi(n-1) + a_n\varphi(n+1) + b_n\varphi(n) = \frac{z + z^{-1}}{2} \varphi(n),$$

(3.9.37)

where $\alpha = z + z^{-1}$ and

$$a_n = \frac{1}{2} e^{-(Q_{n+1} - Q_n)/2}, \qquad b_n = \frac{1}{2} \dot{Q}_n.$$

(3.9.38)

Thus we see that $\varphi(n)$ thus introduced satisfies the equation (3.2.16) for the inverse scattering method. In regard to its time evolution, we first notice the equation,

$$\frac{d}{dt}(Q_n + Q_{n-1} - 2Q'_{n-1})$$

$$= A(e^{-(Q_n' - Q_n)} - e^{-(Q_{n-1}' - Q_{n-1})}) - \frac{1}{A}(e^{-(Q_n - Q_{n-1}')} - e^{-(Q_{n-1} - Q_{n-2}')}) \quad (3.9.39)$$

which is obtained from (3.9.16), and insert from (3.9.36) to obtain ($A = -1$)

$$\frac{\dot{\varphi}(n-1)}{\varphi(n-1)} + a_{n-1}\frac{\varphi(n)}{\varphi(n-1)} - a_{n-2}\frac{\varphi(n-2)}{\varphi(n-1)}$$

$$= \frac{\dot{\varphi}(n)}{\varphi(n)} + a_n\frac{\varphi(n+1)}{\varphi(n)} - a_{n-1}\frac{\varphi(n-1)}{\varphi(n)}. \tag{3.9.40}$$

Therefore we have

$$\dot{\varphi}(n) = a_{n-1}\varphi(n-1) - a_n\varphi(n+1) + F(t)\varphi(n) \tag{3.9.41}$$

where $F(t)$ is a factor independent of n, and reflects the arbitrariness of B in the Lax formalism $L_t = BL - LB$, so that we may put $F(t) \equiv 0$. Thus, we can obtain a new solution Q_n' by first constructing a_n and b_n from a known solution Q_n and then solving for $\varphi(n)$, the equations (3.9.37, 41) by the I.S.M. These equations also give a relationship between the Bäcklund transformation and the I.S.M.

III) We shall briefly describe a dual expression of the Bäcklund transformation for the lattice with exponential interaction [3.9]. The dual system is expressed in terms of $r_n = Q_{n+1} - Q_n$ and its conjugate momentum s_n. Inserting the relation $s_{n-1} - s_n = \dot{Q}_n$ (1.4.7), into (3.9.16), we have (for the sake of simplicity we put $A = 1$)

$$\left.\begin{array}{l} s_n - s_n' = e^{Q_n'(r') - Q_{n+1}(r)} - \lambda \\ s_{n-1}' - s_n = e^{Q_n(r) - Q_n'(r')} - \lambda \end{array}\right\} \tag{3.9.42}$$

In these equations $Q_n(r)$ and $Q_n'(r')$ are to be understood respectively as functions of $\{r_j\}$ and $\{r_j'\}$, or

$$\left.\begin{array}{l} Q_n = Q_n(r) = \sum\limits_{j=-\infty}^{n-1} r_j \\ Q_n' = Q_n'(r') = \sum\limits_{j=-\infty}^{n-1} r_j' \end{array}\right\}, \tag{3.9.43}$$

where we have assumed the dual lattice to be extending from $j = -\infty$, instead of a finite region $j = 0 \sim N$.

The Bäcklund transformation is obtained by making use of the relations (2.2.14), or

$$\left.\begin{array}{l} e^{-(Q_{n+1}(r) - Q_n(r))} - 1 = \dot{s}_n \\ e^{-(Q_{n+1}'(r') - Q_n'(r'))} - 1 = \dot{s}_n' \end{array}\right\}. \tag{3.9.44}$$

We introduce W_n and W_n' by

$$s_n = W_n, \qquad s_{n-1}' = W_n' \tag{3.9.45}$$

and rewrite the left-hand side of (3.9.44) by (3.9.42) to obtain

$$\left. \begin{array}{l} (\lambda + W'_n - W_n)(\lambda + W_n - W'_{n+1}) = 1 + \dot{W}_n \\ (\lambda + W'_n - W_n)(\lambda + W_{n-1} - W'_n) = 1 + \dot{W}'_n \end{array} \right\}.$$ (3.9.46)

This form of the Bäcklund transformation is directly obtained [3.11] without referring to (3.9.16).

Problem 3.9. Put $Q_{n+1} - Q_n = 0$, $\dot{Q}_n = 0$ in (3.9.38) and solve (3.9.37) for $\varphi(n)$, and thus lead the soliton solution (3.9.18a).

3.10 A Finite Lattice

Moser [3.7] discussed a finite lattice of particles $n = 0,1,2, \ldots, N + 1$ with exponential interaction with both end particles at infinity, namely

$$Q_0 = -\infty, \qquad Q_{N+1} = +\infty \qquad \text{(both fixed)}.$$

In this case, since we have pushed both ends to infinity the contribution from the second term ar in the interaction potential becomes infinite. However, as it is constant, we can disregard it. The equations of motion include only the exponential repulsion term $(a/b)\exp(-br)$, so that the system is equivalent to a one-dimensional system of N particles with exponential repulsion between the nearest neighbors.

I) For this system, we number the eigenvalues $\lambda_j(j = 1,2, \ldots, N)$ of $L\varphi = \lambda\varphi$ according to the magnitude, assuming they are all different, in such a way that

$$\lambda_1 < \lambda_2 < \cdots < \lambda_N.$$ (3.10.1)

Intuitively, at $t \to \pm\infty$, all the particles separate out infinitely, and since $a_n = (1/2)\exp(-r_n/2) \to 0$ in this limit, we are left only with the diagonal elements $b_n = (1/2)P_n$ of L. Therefore L becomes asymptotically a diagonal matrix. Thus we have (cf. Fig. 3.3)

$$\left. \begin{array}{ll} P_k = 2\lambda_{N-k+1} & (t \to -\infty) \\ P_k = 2\lambda_k & (t \to +\infty) \end{array} \right\}.$$ (3.10.2)

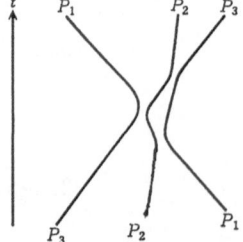

Fig. 3.3. Moser's case

From (3.10.1) we have the boundary conditions

$$a_0 = a_N = 0, \qquad b_0 = b_{N+1} = 0 \tag{3.10.3}$$

so that, as in (3.8.1, 2), the matrices L and B have vanishing elements at both upper right and lower left. Therefore the eigenvalue equations $L\varphi_j = \lambda_j \varphi_j$ and the equations of motion simplify for both $n = 1$ and $n = N$. Especially for $n = 1$, we have

$$a_1 \varphi_j(2, t) + b_1 \varphi_j(1, t) = \lambda_j \varphi (1, t) \tag{3.10.4}$$

$$\dot{\varphi}_j(1, t) = -a_1 \varphi_j(2, t) . \tag{3.10.5}$$

Since $\varphi_j(n, t)$'s for different n are orthogonal, after normalizing we have

$$\sum_{j=1}^{N} \varphi_j(n, t)\varphi_j(n', t) = \delta_{nn'} \tag{3.10.6a}$$

$$\sum_{n=1}^{N} \varphi_j(n, t)\varphi_{j'}(n, t) = \delta_{jj'}. \tag{3.10.6b}$$

Thus (3.10.4) yields

$$\sum_{j=1}^{N} \lambda_j \varphi_j^2(1, t) = b_1 \tag{3.10.7}$$

and rewriting the right-hand side of (3.10.5) by using (3.10.4), we have

$$\dot{\varphi}_j(1, t) = -\left[\lambda_j - \sum_{l=1}^{N} \lambda_l \varphi_l^2(1, t)\right] \varphi_j(1, t) \tag{3.10.8}$$

where we have used the fact that $\dfrac{d}{dt} \sum_{j=1}^{N} \varphi_j^2(1, t) = 0$ by (3.10.6a). Solving (3.10.8) we obtain

$$\varphi_j^2(1, t) = \frac{\varphi_j^2(1, 0) \, e^{-2\lambda_j t}}{\sum\limits_{l=1}^{N} \varphi_l^2(1, 0) \, e^{-2\lambda_l t}} . \tag{3.10.9}$$

In fact, by taking the natural logarithm of (3.10.9) and differentiating we recover (3.10.8).

The system described by $D = \{a_n, b_n\}$ is transferred to

$$\Lambda = [\lambda_1, \cdots, \lambda_N; \varphi_1(1,t), \cdots, \varphi_N(1, t)] \tag{3.10.10}$$

$$\text{with } \sum_{l=1}^{N} |\varphi_l(1, t)|^2 = 1$$

by the mapping $D \to \Lambda$. As we have the conservation of the total momentum

$\sum_n b_n = 1/2 \sum_n P_n = $ const., the dimensions of either space are $2N - 1$. The mapping $A \to D$ implies the inverse problem of $D \to A$.

For $N = 2$, we have $L = \begin{pmatrix} b_1 & a_1 \\ a_1 & b_2 \end{pmatrix}$, so that $L\varphi_j = \lambda_j \varphi_j$ for $j = 1$ reduces to

$$\left.\begin{array}{l} b_1\varphi_1(1, t) + a_1\varphi_1(2, t) = \lambda_1\varphi_1(1, t) \\ a_1\varphi_1(1, t) + b_2\varphi_1(2, t) = \lambda_1\varphi_1(2, t) \end{array}\right\} \qquad (3.10.11)$$

and thus we have

$$-\varphi_1(2, t) = \frac{b_1 - \lambda_1}{a_1} \varphi_1(1, t) = \frac{a_1}{b_2 - \lambda_1} \varphi_1(1, t). \qquad (3.10.12)$$

Also, from (3.10.6b) we have

$$1 = \varphi_1^2(1, t) + \varphi_1^2(2, t)$$
$$= \left(1 + \frac{b_1 - \lambda_1}{b_2 - \lambda_1}\right) \varphi_1^2(1, t). \qquad (3.10.13)$$

If we note the trace formula

$$\operatorname{tr}\{L\} = b_1 + b_2 = \lambda_1 + \lambda_2, \qquad (3.10.14)$$

we obtain

$$\varphi_1^2(2, t) = \frac{b_1 - \lambda_1}{\lambda_2 - \lambda_1}, \qquad \varphi_1^2(1, t) = \frac{b_1 - \lambda_2}{\lambda_1 - \lambda_2}. \qquad (3.10.15)$$

Similarly for $j = 2$, we have

$$\left.\begin{array}{l} \varphi_2^2(2, t) = \dfrac{b_1 - \lambda_2}{\lambda_1 - \lambda_2} = \varphi_1^2(1, t) \\[2mm] \varphi_2^2(1, t) = \dfrac{b_1 - \lambda_1}{\lambda_2 - \lambda_1} = \varphi_1^2(2, t) \end{array}\right\}. \qquad (3.10.16)$$

We solve these for b_1 and b_2, and (3.10.12) for a_1 in terms of $\varphi_1(1, t)$ and $\varphi_2(1, t)$ to obtain

$$\left.\begin{array}{l} b_1 = \lambda_1\varphi_1^2(1, t) + \lambda_2\varphi_2^2(1, t) \\ b_2 = \lambda_2\varphi_1^2(1, t) + \lambda_1\varphi_2^2(1, t) \\ a_1 = (\lambda_2 - \lambda_1)\varphi_1(1, t)\varphi_2(1, t) \end{array}\right\}, \qquad (3.10.17)$$

where we have taken into account that we may assume $\varphi_1(1, t)\,\varphi_2(1, t) > 0$. If

(3.10.9) with $N = 2$ is inserted into (3.10.17), then b_1, b_2 and a_1 are given in terms of the initial conditions $\varphi_1(1, 0)$, $\varphi_2(1, 0)$.

II) For $N > 2$, we need a general representation of the mapping (3.10.10) [3.7]. For this purpose, we denote by λ_k the root of $(L - \lambda I) \varphi = 0$. Then we see that elements of $(L - \lambda I)^{-1}$ can be written as sums of $(\lambda_k - \lambda)^{-1}$. In fact, from

$$L\varphi_k = \lambda_k \varphi_k , \qquad (3.10.18)$$

we have

$$(L - \lambda I)\varphi_k = (\lambda_k - \lambda)\varphi_k; \qquad (3.10.19)$$

so that for

$$R = (L - \lambda I)^{-1} \qquad (3.10.20)$$

we have

$$R\varphi_k = \frac{1}{\lambda_k - \lambda} \varphi_k . \qquad (3.10.21)$$

Therefore, if $f(\lambda)$ is written for the $(1, 1)$ element of R, we have the partial fraction expansion

$$f(\lambda) \equiv R_{11} = \sum_{k=1}^{N} \frac{r_k^2}{\lambda_k - \lambda}, \qquad (3.10.22)$$

where we have chosen a real expression for $\varphi_k(1)$ and have put

$$r_k^2 = \varphi_k^2(1) . \qquad (3.10.23)$$

Letting $|\lambda| \to \infty$, we see that

$$\sum_{k=1}^{N} r_k^2 = 1 . \qquad (3.10.24)$$

If we can express R_{11} as a function of a_n and b_n, (3.10.22, 23) will relate these to $\varphi_k^2(t)$, which is given by (3.10.9), so that the dynamical variables a_n and b_n are explicitly found as functions of time. We have

$$\det R = [\det (L - \lambda I)]^{-1} \qquad (3.10.25)$$

which connects $\det R$ to $\varDelta_N = \det(L - \lambda I)$. We can show that the elements of R, especially R_{11}, are expressible in terms of certain minor determinants of R. To show this, we introduce

$$\Delta_{N-n+1} \equiv \begin{vmatrix} b_n - \lambda & a_n & & & \\ a_n & b_{n+1} - \lambda & & & \\ & & \ddots & & \\ & & & b_{N-1} - \lambda & a_{N-1} \\ & & & a_{N-1} & b_N - \lambda \end{vmatrix} \tag{3.10.26}$$

which is the determinant of a matrix obtained from R by omitting the first $n - 1$ rows and columns. If we expand Δ_N we have

$$\Delta_N = (b_1 - \lambda) \Delta_{N-1} - a_1^2 \Delta_{N-2} \tag{3.10.27}$$

and in general

$$\Delta_{N-n+1} = (b_n - \lambda) \Delta_{N-n} - a_n^2 \Delta_{N-n-1} \qquad (n = 1, 2, \cdots). \tag{3.10.28a}$$

If we define

$$\Delta_{-1} = 0, \qquad \Delta_0 = 1, \tag{3.10.28b}$$

(3.10.28a) holds up to $n = N$.

Now, we introduce an $N \times 1$ matrix with the elements

$$e_1(l) = \sum_{k=1}^N \varphi_k(l)\varphi_k(1) = \delta_{11} \qquad (l = 1, 2, \cdots, N) \tag{3.10.29}$$

and an $N \times 1$ matrix z defined by

$$z(l) = \sum_{k=1}^N \frac{\varphi_k(l)\varphi_k(1)}{\lambda_k - \lambda}. \tag{3.10.30}$$

Then (3.10.21) yields

$$Re_1 = z. \tag{3.10.31}$$

By (3.10.19) we may rewrite it as

$$(L - \lambda I) z = e_1. \tag{3.10.32}$$

Writing down each element, we thus obtain a set of simultaneous equations

$$\left.\begin{aligned} &(b_1 - \lambda)z(1) + a_1 z(2) = 1 \\ &a_1 z(1) + (b_2 - \lambda)z(2) + a_2 z(3) = 0 \\ &\qquad\cdots\cdots\cdots\cdots\cdots\cdots \\ &a_{N-1} z(N - 1) + (b_N - \lambda)z(N) = 0 \end{aligned}\right\}. \tag{3.10.33}$$

The determinant of the coefficients of $z(n)$ is Δ_N itself, and by (3.10.30) $z(1)$ is nothing but R_{11}. Thus, solving (3.10.33) for $z(1)$ we obtain

$$f(\lambda) = R_{11} = z(1) = \frac{\Delta_{N-1}}{\Delta_N}. \tag{3.10.34}$$

Rewriting (3.10.27), we have

$$\frac{\Delta_N}{\Delta_{N-1}} = b_1 - \lambda - \frac{a_1^2}{\dfrac{\Delta_{N-1}}{\Delta_{N-2}}} \tag{3.10.35a}$$

and similarly from (3.10.26) we obtain

$$\frac{\Delta_{N-n+1}}{\Delta_{N-n}} = b_n - \lambda - \frac{a_n^2}{\dfrac{\Delta_{N-n}}{\Delta_{N-n-1}}}. \tag{3.10.35b}$$

Therefore by successive reduction, we have a finite continued fraction for $f(\lambda) = R_{11}$, which states that [3.7]

$$f(\lambda) = \cfrac{1}{b_1 - \lambda - \cfrac{a_1^2}{b_2 - \lambda \atop \ddots \atop \displaystyle - \cfrac{a_{N-1}^2}{b_N - \lambda}}} \tag{3.10.36}$$

Thus we have two expressions, a partial fraction (3.10.22) and a continued fraction (3.10.23) for the same $f(\lambda)$. Identifying these we can express b_n and a_n in terms of r_k^2 following a method described by *Stieltjes* (cf. Appendix A) [3.12]. The time dependence of $r_k^2 = r_k^2(t)$ is given by (3.10.9, 23). Therefore we have

$$\left. \begin{aligned} b_n &= b_n(\{r_k(0)\}, \{\lambda_k\}) \\ a_n &= a_n(\{r_k(0)\}, \{\lambda_k\}) \end{aligned} \right\} \tag{3.10.37}$$

which give the solution to the inverse problem of the mapping (3.10.10) (for the concrete expression, see Appendix A), and provide another I.S.M. to the lattice with exponential interaction.

III) We can derive the continued fraction (3.10.36) and the time rate of change of the partial fraction (3.10.22) without referring to the Lax formalism, but using the conserved quantities of Hénon's type, and thus integrating the motion by the aid of these quantities [3.13]. First we put

$$\left. \begin{aligned} A_j &= e^{-(Q_{j+1}-Q_j)} = (2a_j)^2 \\ B_j &= P_j = 2b_j \end{aligned} \right\}; \tag{3.10.38}$$

then the equations of motion for the lattice take the form

$$\frac{dA_j}{dt} = A_j(B_j - B_{j+1}), \qquad \frac{dB_j}{dt} = A_{j-1} - A_j . \tag{3.10.39}$$

On the other hand, using the conserved quantities (2.10.19) of Hénon's type, i.e.,

$$I_n = \left(\sum_{j=1}^{N} \frac{\partial}{\partial B_j}\right)^{N-n} \exp\left(-\sum_{k=1}^{N-1} A_k \frac{\partial^2}{\partial B_k \partial B_{k+1}}\right) \prod_{l=1}^{N} B_l , \tag{3.10.40}$$

we make a conserved quantity

$$f_1(\Lambda) = \sum_{n=0}^{N} \frac{(-\Lambda)^{N-n}}{(N-n)!} I_n = \exp\left(-\sum_{k=1}^{N-1} A_k \frac{\partial^2}{\partial B_k \partial B_{k+1}}\right) \prod_{l=1}^{N} (B_l - \Lambda) . \tag{3.10.41}$$

Expanding the right-hand side of (3.10.41) we see that

$$f_1(\Lambda) \propto \det (2L - \Lambda I) . \tag{3.10.42}$$

Thus the roots of $f_1(\Lambda) = 0$ are twice the $\lambda_j (j = 1,2, \ldots, N)$ we have already introduced. That is $\Lambda_j = 2\lambda_j$. Further defining

$$f_k(\Lambda) = \exp\left(-\sum_{j=k}^{N-1} A_j \frac{\partial^2}{\partial B_j \partial B_{j+1}}\right) \prod_{l=k}^{N} (B_l - \Lambda) , \tag{3.10.43}$$

we see that

$$f_k(\Lambda) = (B_k - \Lambda) f_{k+1}(\Lambda) - A_k f_{k+2}(\Lambda) \tag{3.10.44a}$$

holds. Now, instead of a periodic lattice, we assume as Moser did, that end particles are at infinity ($Q_0 = -\infty$, $Q_{N+1} = +\infty$, so that $A_0 = A_N = 0$) and thus the end particles do not interfere with the motion in the lattice. We can also use Hénon's conserved quantities under this condition. Further we may define

$$f_N(\Lambda) = B_N - \Lambda, \qquad f_{N+1}(\Lambda) = 1 . \tag{3.10.44b}$$

Since $A_k > 0$ ($k = 1,2, \ldots, N - 1$) in (3.10.44a), the roots of $f_k(\Lambda) = 0$ constitute "suite de Sturm" (cf. Appendix B), and so each of them is between the roots of $f_{k+1}(\Lambda) = 0$. Therefore at the root Λ_j of $f_1(\Lambda)$, we have $f_k(\Lambda_j) \neq 0$ ($k \neq 1$).

From (3.10.44a) we have a continued fraction expansion

$$\frac{f_2(\Lambda)}{f_1(\Lambda)} = \cfrac{1}{B_1 - \Lambda - \cfrac{A_1}{B_2 - \Lambda - \cfrac{\ddots}{-\cfrac{A_{N-1}}{B_N - \Lambda}}}} . \tag{3.10.45}$$

Especially for $k = 1$, (3.10.44a) reduces to

$$f_1(\Lambda) = (B_1 - \Lambda)f_2(\Lambda) - A_1 f_3(\Lambda) \tag{3.10.46}$$

and comparing (3.10.27) with (3.10.46, 42) we see that

$$\frac{f_2(\Lambda)}{f_1(\Lambda)} = \frac{\Delta_{N-1}}{2\Delta_N}. \tag{3.10.47}$$

Thus we find that the continued fractions (3.10.45 and 36) are equivalent. From (3.10.46) we have

$$\frac{\partial}{\partial B_1} f_1(\Lambda) = f_2(\Lambda) \tag{3.10.48}$$

and with the roots Λ_2 of $f_1(\Lambda)$ we may write

$$f_1(\Lambda) = \prod_{l=1}^{N} (\Lambda_l - \Lambda) \tag{3.10.49}$$

so that we obtain

$$f_2(\Lambda) = \sum_{j=1}^{N} \frac{\partial \Lambda_j}{\partial B_1} \prod_{l}^{(j)} (\Lambda_l - \Lambda), \tag{3.10.50}$$

where $\prod^{(j)}$ stands for the product omitting the factor $\Lambda_j - \Lambda$. If we put $\Lambda = \Lambda_j$ in (3.10.50), we have

$$\frac{\partial \Lambda_j}{\partial B_1} = \frac{f_2(\Lambda_j)}{\prod_{l}^{(j)}(\Lambda_l - \Lambda_j)}. \tag{3.10.51}$$

Thus, from (3.10.49–51), we have the partial fraction expansion

$$\frac{f_2(\Lambda)}{f_1(\Lambda)} = -\sum_{j=1}^{N} \frac{\partial \Lambda_j/\partial B_1}{\Lambda - \Lambda_j}$$

$$= -\sum_{j=1}^{N} \frac{1}{\Lambda - \Lambda_j} \frac{f_2(\Lambda_j)}{\prod_{l}^{(j)}(\Lambda_l - \Lambda_j)}. \tag{3.10.52}$$

Further, putting $\Lambda = \Lambda_j$ in (3.10.46) we have

$$\left(\frac{f_3}{f_2}\right)_j = \frac{B_1 - \Lambda_j}{A_1} \tag{3.10.53}$$

where, and in what follows, we use the general abbreviation

$$f_{k+1}(\Lambda_j)/f_k(\Lambda_j) = (f_{k+1}/f_k)_j.$$

If we differentiate (3.10.53) with respect to time, by using (3.10.39) and the boundary condition $A_0 = 0$, we obtain

$$\frac{d}{dt}\left(\frac{f_3}{f_2}\right)_j = -1 - (B_1 - B_2)\left(\frac{f_3}{f_2}\right)_j .$$
(3.10.54)

In general we have

$$\frac{d}{dt}\left(\frac{f_{k+2}}{f_{k+1}}\right)_j = -1 - (B_k - B_{k+1})\left(\frac{f_{k+2}}{f_{k+1}}\right)_j + \frac{(f_{k+2}/f_{k+1})_j}{(f_{k+1}/f_k)_j} ,$$
(3.10.55)

though we skip the proof which can be achieved by mathematical induction. Rewriting this as

$$\frac{d}{dt} \ln \left(\frac{f_{k+2}}{f_{k+1}}\right)_j = -\frac{d}{dt} \ln A_k - \frac{1}{(f_{k+2}/f_{k+1})_j} + \frac{1}{(f_{k+1}/f_k)_j} ,$$
(3.10.56)

we add together from $j = 1$ to N to obtain

$$\frac{d}{dt} \ln \frac{f_2(\Lambda_j)}{A_1 A_2 \cdots A_{N+1}} = f_N(\Lambda_j) = B_N - \Lambda_j ,$$
(3.10.57)

where we have taken into account $f_1(\Lambda_j) = 0, f_{N+1}(\Lambda_j) = 1$ and $f_N(\Lambda_j) = B_N - \lambda_j$ by (3.10.44a). Therefore we have

$$\frac{d}{dt} \ln \frac{f_2(\Lambda_j)}{f_2(\Lambda_k)} = \Lambda_k - \Lambda_j .$$
(3.10.58)

Thus, introducing $F = F(t)$ which is independent of j, and constants C_j which are determined by the initial conditions, we see that (3.10.58) is satisfied by

$$f_2(\Lambda_j) = C_j e^{-\Lambda_j t}/F .$$
(3.10.59)

To determine F we note

$$I_1 = \sum_{j=1}^{N} \Lambda_j = \sum_{j=1}^{N} B_j .$$
(3.10.60)

Then, by (3.10.51) we have

$$1 = \frac{\partial I_1}{\partial B_1} = \sum_{j=1}^{N} \frac{\partial \Lambda_j}{\partial B_1} = \sum_{j=1}^{N} \frac{f_2(\Lambda_j)}{\prod_l^{(j)}(\Lambda_l - \Lambda_j)}$$
(3.10.61)

and inserting from (3.10.59), we finally obtain

$$F = \sum_{s=1}^{N} \frac{C_s e^{-\Lambda_s t}}{\prod_l^{(s)}(\Lambda_l - \Lambda_s)} .$$
(3.10.62)

If we insert the last expression and (3.10.59) into (3.10.52) we obtain $f_2(\Lambda)/f_1(\Lambda)$ as a function of time expressed as a partial fraction expansion. On the other hand, we have the same $f_2(\Lambda)/f_1(\Lambda)$ expressed as a continued fraction expansion (3.10.45). Equating these, we can solve the coefficients A_j and B_j in the continued fraction by the use of Stieltjes' method (Appendix A) as functions of time [3.14]. The above is probably the most direct method to the solution using the conserved quantities. The same idea may be applicable to a periodic lattice, though has not succeeded yet.

Problem 3.10. Equating (3.10.36, 22), derive (3.10.17) for the case $N = 2$.

Problem 3.11. By using (3.10.54), prove (3.10.55) by mathematical induction.

Hint: After rewriting (3.10.44a) as

$$A_l \left(\frac{f_{l+2}}{f_{l+1}} \right)_j = B_j - A_j - \frac{1}{(f_{l+1}/f_l)_j},$$

differentiate both sides with respect to t, and assume, on the right-hand side, (3.10.55) with $k = l - 1$ to hold.

3.11 Continuum Approximation

For waves with small variation compared with the average distance between particles in the lattice, that is for smooth waves with long wavelength, we can use the continuum approximation. We will see that the equations of motion for the lattice reduce to the KdV equation if we retain the lowest nonlinear term, and the I.S.M. for the lattice reduces to that for the KdV equation [3.5b, 15, 16].

I) For the matrices L and B, we have

$$\left. \begin{array}{l} (L\varphi)_n = a_{n-1}\varphi(n-1) + b_n\varphi(n) + a_n\varphi(n+1) \\ (B\varphi)_n = a_{n-1}\varphi(n-1) - a_n\varphi(n+1) \end{array} \right\}. \tag{3.11.1}$$

Therefore if we define shifting operators by

$$e^{\pm \partial/\partial n} f(n) = f(n \pm 1) \tag{3.11.2}$$

and treat n as a continuous variable, we may write

$$\left. \begin{array}{l} L = b_n + e^{-\partial/\partial n} a_n + a_n e^{\partial/\partial n} \\ B = e^{-\partial/\partial n} a_n - a_n e^{\partial/\partial n} \end{array} \right\}. \tag{3.11.3}$$

Further, if we use formal expansions

$$
\left.
\begin{aligned}
e^{\pm \partial/\partial n} &= 1 \pm \frac{\partial}{\partial n} + \frac{\partial^2}{\partial n^2} \pm \frac{1}{6} \frac{\partial^4}{\partial n^3} + \cdots \\
a_n &= \frac{1}{2} e^{-r_n/2} = \frac{1}{2} \left(1 - \frac{r_n}{2} \right)
\end{aligned}
\right\}
$$

(3.11.4)

and $b_n = P_n/2$, we obtain

$$
\left.
\begin{aligned}
L &= \frac{1}{2} P_n + \left(1 - \frac{r_n}{2} \right) + \frac{1}{4} \left(\frac{\partial}{\partial n} r_n - r_n \frac{\partial}{\partial n} \right) + \frac{1}{2} \frac{\partial^2}{\partial n^2} \\
B &= -\frac{\partial}{\partial n} + \frac{1}{4} \left(\frac{\partial}{\partial n} r_n + r_n \frac{\partial}{\partial n} \right) - \frac{1}{6} \frac{\partial^3}{\partial n^3}
\end{aligned}
\right\}.
$$

(3.11.5)

In what follows, we focus our attention to waves propagating to the right. We introduce a coordinate frame ξ which moves to the right with the velocity for infinitely small waves ($c_0 = 1$), a rescaled time τ, and r_n with the new variables (ξ, τ), in such a way that

$$
\xi = n - t, \qquad \tau = t/24, \qquad u = 2r_n .
$$

(3.11.6)

Now, $\partial \varphi/\partial t = B\varphi$ gives

$$
\frac{\partial}{\partial \tau} \varphi = \tilde{B}\varphi ,
$$

(3.11.7)

where $\tilde{B}$ is a rescaled B defined as

$$
\tilde{B} = -4 \frac{\partial^3}{\partial \xi^3} + 3 \left(\frac{\partial}{\partial \xi} u + u \frac{\partial}{\partial \xi} \right) .
$$

(3.11.8)

Thus we see that $\partial/\partial \tau$ consists of higher derivatives or higher small quantities, so that we may use

$$
P_n = -\frac{\partial Q}{\partial \xi} + \frac{1}{24} \frac{\partial Q}{\partial \tau} \simeq -r_n = -\frac{u}{2}
$$

(3.11.9)

in this approximation. We can also neglect the third term of L in (3.11.5) as a small quantity. Thus using (3.11.9) we have

$$
L = \frac{1}{2} \frac{\partial^2}{\partial \xi^2} - \frac{u}{2} + 1
$$

(3.11.10)

and the equation of motion

$$
L_\tau = \tilde{B}L - L\tilde{B}
$$

(3.11.11)

gives the KdV equation (subscripts mean derivatives)

$$u_\tau - 6uu_\xi + u_{\xi\xi\xi} = 0 . \tag{3.11.12}$$

When n and m are taken as continuous variables in the continuum approximation, we use primes to indicate the corresponding quantities. Thus for (3.4.19, 20) we write

$$\kappa'(n, m) = \kappa'(n, m)h, \qquad F(n + m) = F'(n + m)h , \tag{3.11.13}$$

where h stands for the average distance between the particles in the lattice, and we may write $h = dn$. In this approximation, the GL equation (3.4.21) reduces to an integral equation

$$\kappa'(n, m) + F'(n + m) + \int_n^\infty \kappa'n, n')F(n' + m) \, dn' = 0 \tag{3.11.14}$$

and (3.4.22) to

$$\frac{1}{[K(n, n)]^2} = 1 + F'(2n) + \int_n^\infty \kappa'(n, n')F'(n' + n)dn' . \tag{3.11.15}$$

If we further write

$$\left.\begin{array}{ll} z = e^{ik}, & z_J = e^{-k_J} \\ n - t = \xi, & m - t = \eta \\ F'(n + m) = \tilde{F}(\xi + \eta), & \kappa'(n, m) = \tilde{\kappa}(\xi, \eta) \end{array}\right\}, \tag{3.11.16}$$

the kernel (3.4.19) reduces to

$$\tilde{F}(\xi + \eta; \tau) = \tilde{F}(\xi + \eta) = \frac{1}{2\pi} \int_{-\infty}^\infty R(k, 0) \, e^{8ik^3\tau + ik(\xi+\eta)} \, dk$$

$$+ \sum c_j^2(0) \, e^{8k_J^3\tau - k_J(\xi+\eta)} \tag{3.11.17}$$

when higher order terms are neglected. Then the GL equation reduces to

$$\tilde{\kappa}(\xi, \eta) + \tilde{F}(\xi + \eta) + \int_\xi^\infty \tilde{\kappa}(\xi, \zeta) \, \tilde{F}(\zeta + \eta) \, d\zeta = 0 . \tag{3.11.18}$$

Further, (3.11.15) gives

$$K(\xi, \xi)^{-2} = 1 + \tilde{F}(2\xi) + \int_\xi^\infty \tilde{\kappa}(\xi, \zeta) \, \tilde{F}(\zeta + \xi) \, d\zeta . \tag{3.11.19}$$

However, if we subtract (3.11.18) with $\eta \to \xi$ from (3.11.19) we have

$$K(\xi, \xi)^{-2} = 1 - \tilde{\kappa}(\xi, \xi) \tag{3.11.20}$$

and therefore from (3.6.26)

$$e^{-r_n} = \left[\frac{K(n, n)}{K(n-1, n-1)} \right]^2$$

$$\simeq \frac{1 - \bar{\kappa}(\xi - 1, \xi - 1)}{1 - \bar{\kappa}(\xi, \xi)}$$

$$\simeq 1 + [\bar{\kappa}(\xi, \xi) - \bar{\kappa}(\xi - 1, \xi - 1)] . \tag{3.11.21}$$

Thus, in this approximation, $u = 2r_n$ is given by

$$u(\xi, t) = -2 \frac{\partial}{\partial \xi} \bar{\kappa}(\xi, \xi) . \tag{3.11.22}$$

(3.11.17, 18) and (3.11.22) are well known equations of the I.S.M. for the KdV equation [3.17].

II) The Schrödinger equation $L\varphi = \lambda\varphi$ associated with the KdV equation is usually written as

$$\frac{d^2\varphi}{d\xi^2} + (\lambda^{(\mathrm{KdV})} - u)\,\varphi = 0, \qquad \lambda^{(\mathrm{KdV})} = k^2 . \tag{3.11.23}$$

Comparing this with (3.11.10), or

$$\left. \begin{array}{l} \left(\dfrac{1}{2} \dfrac{d^2\varphi}{d\xi^2} - \dfrac{u}{2} + 1 \right) \varphi = \lambda\varphi \\[2mm] \lambda = \dfrac{z + z^{-1}}{2} = \cos k \end{array} \right\}, \tag{3.11.24}$$

one finds a relationship between the spectrum λ (for example, of solitons) of the waves propagating to the right in the lattice and the spectrum $\lambda^{(\mathrm{KdV})}$ of the KdV equation. For $\lambda \simeq 1$ this is given as

$$\lambda = \cos \sqrt{\lambda^{(\mathrm{KdV})}} \tag{3.11.25}$$

or, since $\lambda^{(\mathrm{KdV})}$ is assumed small,

$$\lambda \simeq 1 - \frac{\lambda^{(\mathrm{KdV})}}{2} . \tag{3.11.26}$$

We have thus obtained the KdV equation as a continuum approximation for waves propagating to the right. As already remarked, we may discuss the lattice with exponential interaction similarly when we change the sign of a_n. With this modification the similar continuum approximation leads to the KdV equation

$$u_\tau + 6uu_\xi - u_{\xi\xi\xi} = 0 \tag{3.11.27}$$

which is for waves propagating to the left and has its spectrum near $\lambda \simeq -1$.

III) We will briefly describe a continuum approximation of the Bäcklund transformation [3.16]. We first note the fact that the time derivative implies

$$\frac{\partial}{\partial t} = -\frac{\partial}{\partial n} + \frac{1}{24}\frac{\partial}{\partial \tau}.$$

Then we expand Q'_n around n, and Q_n around $n + 1/2$ in (3.9.16) with $A = -1$, to obtain

$$\frac{\partial Q'_n}{\partial n} - \frac{1}{24}\frac{\partial Q'_n}{\partial \tau} = 2 - \alpha - \frac{\partial Q_{n+1/2}}{\partial n} + (Q'_n - Q_{n+1/2})^2$$

$$- \frac{1}{24}\frac{\partial^3 Q_{n+1/2}}{\partial n^3} - \frac{1}{4}(Q'_n - Q_{n+1/2})\frac{\partial^2 Q_{n+1/2}}{\partial n^2}$$

$$+ \frac{1}{4}\left(\frac{\partial Q_{n+1/2}}{\partial n}\right)^2 - \frac{1}{2}(Q'_n - Q_{n+1/})^2\frac{\partial Q_{n+1/2}}{\partial n}, \qquad (3.11.28)$$

and

$$\frac{\partial Q_{n+1/2}}{\partial n} - \frac{1}{24}\frac{\partial Q_{n+1/2}}{\partial \tau} = 2 - \alpha - \frac{\partial Q'_n}{\partial n} + (Q'_n - Q_{n+1/2})^2$$

$$- \frac{1}{24}\frac{\partial^3 Q'}{\partial n^3} + \frac{1}{4}(Q'_n - Q_{n+1/2})\frac{\partial^2 Q'_n}{\partial n^2}$$

$$+ \frac{1}{4}\left(\frac{\partial Q'_n}{\partial n}\right)^2 - \frac{1}{2}(Q'_n - Q_{n+1/2})^2\frac{\partial Q'_n}{\partial n} \qquad (3.11.29)$$

where we have assumed the order of magnitude

$$Q \sim \varepsilon, \qquad \partial/\partial n \sim \varepsilon, \qquad \partial/\partial \tau \sim \varepsilon^3 \qquad\qquad (3.11.30)$$

with small ε, and stopped the expansion at the fourth-order terms. Then, the above equations split into ε^2 order and ε^4 order equations. If we write

$$\left.\begin{array}{l} 2Q_{n+1/2} = w(x, \tau) \\ 2Q'_n = w'(x, \tau) \end{array}\right\} \qquad\qquad (3.11.31)$$

with x instead of n, the ε^2 terms for (3.11.28, 29) give the same equation $2(2 - \infty)$

$$w'_x + w_x = 2(2 - \alpha) + (w' - w)^2/2 \qquad\qquad (3.11.32)$$

where the subscripts mean derivatives. As for the ε^4 terms, if we take the difference of (3.11.28, 29) and use the time derivative of (3.11.32), we obtain

$$w_\tau - w'_\tau = -(w_{xxx} - w'_{xxx}) + 3(w_x^2 - w_x'^2). \qquad\qquad (3.11.33)$$

Equations (3.11.32, 33) constitute the Bäcklund transformation [3.18] for the KdV equation written in the form

$$w_\tau - 3w_x^2 + w_{xxx} = 0 \qquad (u = w_x). \qquad\qquad (3.11.34)$$

4. Periodic Systems

In this chapter we discuss periodic lattices.[1] As we have seen in the last chapter, the I.S.M. for an infinite lattice makes use of the discrete Schrödinger equation. For periodic systems this gives a discrete Hill's equation, and in place of the scattering data, it is convenient to use the spectrum of the discrete Hill's equation and the auxiliary spectrum for fixed boundary conditions of the same equation. In this case the fundamental solutions and the discriminant of the discrete Hill's equation play important roles. The discriminant is a polynomial of the spectrum, and the integral of motion is given in terms of elliptic integrals. Thus the initial value problem reduces to the inverse problem (Jacobi's inverse problem), or inverse spectral theory. To detcribe this problem, we discuss the spectrum on a complex surface, and the integrals on a closed Riemann surface. As a simple example, we discuss again the case of a cnoidal wave. Finally, we show some concrete calculations for a three-particle system, and find some relationship between the discriminant and the auxiliary spectrum, considered as dynamical variables, in order to prepare for the next chapter.

4.1 Discrete Hill's Equation

In this chapter we discuss the initial value problem of a periodic lattice with exponential interaction between neighbors. We assume no impurity, and a system composed of N particles. For dynamical quantities we use the same notations $a_n = a_n(t)$, $b_n = b_n(t)$ etc. as before. The periodic conditions are expressed as

$$a_{n+N} = a_n, \qquad b_{n+N} = b_n . \tag{4.1.1}$$

It is convenient to consider an infinite system

$$- \infty < n < \infty \tag{4.1.2}$$

composed of such a_n and b_n. For this infinite system we discuss the equation

$$(L\varphi)_n \equiv a_{n-1}\varphi(n-1) + b_n\varphi(n) + a_n\varphi(n+1) = \lambda\varphi(n) \tag{4.1.3a}$$

with a constant λ. Since the coefficients a_n and b_n are periodic, the above is a discrete version of Hill's equation [4.1]

---

[1] Part of this Chapter as well as of the following one, are included in[4.7]

$$d^2\varphi/dx^2 + (\lambda - u)\varphi = 0, \qquad u(x + L) = u(x), \tag{4.1.3b}$$

and (4.1.3a) is called the discrete Hill's equation. This is a difference equation of the second rank, which can be solved for $\varphi(n)$ when, for example, the values $\varphi(0)$, and $\varphi(1)$, at $n = 0$ and $n = 1$ respectively, are given. $\varphi(n)$ thus obtained is not periodic in general and may diverge. A solution is said to be stable if it is nondivergent. It will be shown below that stable solutions form bands of the spectrum λ, and between these stable regions there are unstable regions (gaps). Such a spectral structure depends on a_n and b_n, and conversely the spectral structure restricts a_n and b_n to some extent. Though the problem of determining a_n and b_n from the knowledge of the spectrum is an inverse problem, we have no asymptotic solutions at infinity, or scattering data, for a periodic case; it will be more appropriate to call it the inverse spectral problem rather than the inverse scattering problem. It will be shown that if a certain set of data regarding the initial conditions is given we can determine the future evolution of the lattice. The data we need in this case consist of the time-independent spectrum and the initial auxiliary spectrum. Since the inverse problem is more difficult for a periodic system than an infinite one, the whole of the present chapter is devoted to this problem. To begin with, we describe certain properties of the discrete Hill's equation [4.2, 3].

Since the discrete Hill's equation (4.1.3a) is a difference equation of the second rank for a fixed λ, we have two linearly independent solutions (fundamental solutions), and arbitrary solutions for the same λ can be written as a linear combination of these fundamental solutions. Let us denote the fundamental solution $(- \infty < n < \infty)$ of (4.1.3a) by $\varphi_1(n) = \varphi_1(n, \lambda)$ and $\varphi_2(n) = \varphi_2 n, \lambda)$. An arbitrary solution $\varphi(n) = \varphi(n, \lambda)$ can be written as

$$\varphi(n) = c_1\varphi_1(n) + c_2\varphi_2(n). \tag{4.1.4}$$

Though the following results hold irrespective of the choice of the fundamental solutions, for preciseness we use the fundamental solutions defined by the boundary conditions

$$\left. \begin{array}{ll} \varphi_1(0) = 1, & \varphi_1(1) = 0 \\ \varphi_2(0) = 0, & \varphi_2(1) = 1 \end{array} \right\}. \tag{4.1.5}$$

When these are stable, we may say that $\varphi_1(n)$ and $\varphi_2(n)$ are, respectively, cosinelike and sinelike functions. Writing down (4.1.3) for $\varphi_1(n)$, we have

$$\left. \begin{array}{l} a_0 + a_1\varphi_1(2) = 0 \\ b_2\varphi_1(2) + a_2\varphi_1(3) = \lambda\varphi_1(2) \\ a_2\varphi_1(2) + b_3\varphi_1(3) + a_3\varphi_1(4) = \lambda\varphi_1(3) \\ \cdots\cdots\cdots\cdots\cdots \end{array} \right\}, \tag{4.1.6}$$

and solving successively we obtain

$$
\left.
\begin{aligned}
\varphi_1(2) &= -\frac{a_0}{a_1} \\[2mm]
\varphi_1(3) &= -\frac{a_0}{a_1 a_2}(\lambda - b_2) \\[2mm]
\varphi_1(4) &= -\frac{a_0}{a_1 a_2 a_3}[\lambda^2 - (b_2 + b_3)\lambda - a_0 a_1 a_2 + b_2 b_3]
\end{aligned}
\right\}.
$$

$$\dotfill$$

(4.1.7)

In general, for $n \geq 2$ we have

$$
\varphi_1(n) = -a_0 \left(\prod_{j=1}^{n-1} a_j\right)^{-1}\left[\lambda^{n-2} - \left(\sum_{j=2}^{n-1} b_j\right)\lambda^{n-3} + \cdots\right]. \tag{4.1.8}
$$

Similarly, for $n \geq 2$ we have

$$
\varphi_2(n) = \left(\prod_{j=1}^{n-1} a_j\right)^{-1}\left[\lambda^{n-1} - \left(\sum_{j=1}^{n-1} b_j\right)\lambda^{n-2} + \cdots\right]. \tag{4.1.9}
$$

Now, replacing n in (4.1.3a) by $n + N$, and remembering (4.1.1), we see that $\varphi_1(n + N)$ and $\varphi_2(n + N)$ also satisfy (4.1.3a), so that they can be expressed as linear combinations of $\varphi_1(n)$ and $\varphi_2(n)$. We may thus write

$$
\begin{pmatrix} \varphi_1(N + n) \\ \varphi_2(N + n) \end{pmatrix} = M \begin{pmatrix} \varphi_1(n) \\ \varphi_2(n) \end{pmatrix} \tag{4.1.10}
$$

where M is a 2×2 matrix. Putting $n = 0, 1$, and using (4.1.5) we can determine the elements of M as

$$
M = \begin{pmatrix} \varphi_1(N) & \varphi_1(N + 1) \\ \varphi_2(N) & \varphi_2(N + 1) \end{pmatrix}, \tag{4.1.11}
$$

where M is called a monodromy matrix.

On the other hand, φ_1 and φ_2 satisfy

$$
\left.
\begin{aligned}
a_n \varphi_1(n + 1) + b_n \varphi_1(n) + a_{n-1}\varphi_1(n - 1) &= \lambda \varphi_1(n) \\
a_n \varphi_2(n + 1) + b_n \varphi_2(n) + a_{n-1}\varphi_2(n - 1) &= \lambda \varphi_2(n)
\end{aligned}
\right\} \tag{4.1.12}
$$

from which, eliminating λ, we have

$$
\begin{aligned}
W &\equiv a_n[\varphi_1(n)\varphi_2(n + 1) - \varphi_1(n + 1)\varphi_2(n)] \\
&= a_{n-1}[\varphi_1(n - 1)\varphi_2(n) - \varphi_1(n)\varphi_2(n - 1)]. \tag{4.1.13}
\end{aligned}
$$

This is a relationship concerning a discrete version of the Wronskian W for the differences $\varphi_1(n + 1) - \varphi_1(n)$ and $\varphi_2(n + 1) - \varphi_2(n)$. Lifting n to N on one side, and lowering n to 1 on the other side, we obtain

$$
\begin{aligned}
W &= a_N[\varphi_1(N)\varphi_2(N + 1) - \varphi_1(N + 1)\varphi_2(N)] \\
&= a_0[\varphi_1(0)\varphi_2(1) - \varphi_1(1)\varphi_2(0)] \\
&= a_0 ,
\end{aligned} \tag{4.1.14}
$$

where we have used (4.1.5). Further, since $a_0 = a_N$ we have

$$
\det M = \varphi_1(N)\varphi_2(N + 1) - \varphi_1(N + 1)\varphi_2(N) = 1 . \tag{4.1.15}
$$

For some very special values of λ, the solution $\varphi(n)$ of (4.1.4) can be periodic, but more generally we have solutions satisfying

$$
\varphi(n + N) = \rho\varphi(n) , \tag{4.1.16}
$$

which means for $n = 0,1$ that

$$
\left.\begin{aligned}
c_1\varphi_1(N) + c_2\varphi_2(N) &= \rho c_1 \\
c_1\varphi_1(N + 1) + c_2\varphi_2(N + 1) &= \rho c_2
\end{aligned}\right\} . \tag{4.1.17}
$$

Therefore, ρ is the root of the equation

$$
\rho^2 - \Delta(\lambda)\rho + 1 = 0 \tag{4.1.18}
$$

with

$$
\Delta(\lambda) = \varphi_1(N) + \varphi_2(N + 1) = \mathrm{tr}\,\{M\} \tag{4.1.19}
$$

which is called the discriminant. $\varphi(n)$ defined by (4.1.16) is called the Bloch function (originally this was a term for the wave function of an electron moving in a periodic electric field inside a solid). The fact that such a function exists is known as the Floquet theorem. Solving (4.1.18) we have

$$
\rho = \frac{1}{2}(\Delta \pm \sqrt{\Delta^2 - 4}) . \tag{4.1.20}
$$

Equation (4.1.19), $\Delta(\lambda) = \mathrm{tr}\,\{M\}$, holds irrespective of the choice of fundamental solutions.

As a simple example, consider a lattice at rest,

$$
a_n = \frac{1}{2}, \qquad b_n = 0 . \tag{4.1.21}
$$

Then (4.1.3a) reduces to

$$\frac{1}{2}\left[\varphi(n+1) + \varphi(n-1)\right] = \lambda\varphi(n) \qquad (4.1.22)$$

and the solution is obtained by assuming $\varphi \sim \exp(\pm i\alpha n)$. The results is

$$\left.\begin{array}{l} \varphi_1(n) = -\dfrac{\sin \alpha(n-1)}{\sin \alpha} \\[2ex] \varphi_2(n) = \dfrac{\sin \alpha n}{\sin \alpha} \\[2ex] \lambda = \cos \alpha \end{array}\right\}. \qquad (4.1.23)$$

The discriminant is

$$\begin{aligned} \Delta(\lambda) &= \frac{1}{\sin \alpha}\left[-\sin \alpha(N-1) + \sin \alpha(N+1)\right] \\ &= 2\cos \alpha N \end{aligned} \qquad (4.1.24)$$

in this case, and it oscillates between ± 2 for $|\lambda| \leq 1$. The roots λ_j of $\Delta^2 - 4 = 0$ ($\alpha_j = \pi/N$, $j[=0,1,2, \ldots, N)$ are double except $j = 0, N$. For $|\lambda| > 1$, the root α is imaginary, and $|\Delta(\lambda)| \to \infty$ for $\lambda \to \pm \infty$ (Fig. 4.1).

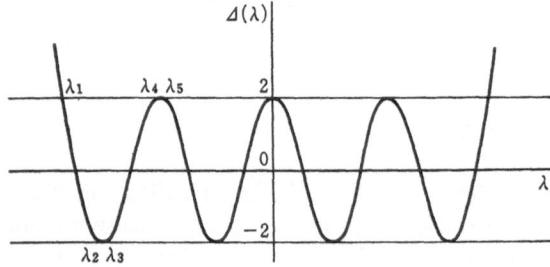

Fig. 4.1. Schematic diagram of $\Delta(\lambda) \sim \lambda$ (without motion). $n =$ even, $g = 0$ (number of simple roots is $2g + 2$)

When λ satisfies

$$\Delta^2(\lambda) \leq 4, \qquad (4.1.25)$$

ρ of (4.1.20) is in general complex, and $|\rho| = 1$. Thus such λ belongs to a stable region, that is, the solution is stable. When $\rho = 1$, the period of the solution is $N[\varphi(n + N) = \varphi(n)]$, and when $\rho = -1$ the period is $2N[\varphi(n + N) = -\varphi(n)]$. When

$$\Delta^2(\lambda) > 4, \qquad (4.1.26)$$

λ belongs to an unstable region. The roots of the equations

$$\Delta(\lambda) - 2 = 0 \tag{4.1.27a}$$

and

$$\Delta(\lambda) + 2 = 0 \tag{4.1.27b}$$

belong to eigenfunctions with periods N and $2N$. It is easy to show that these are, respectively, eigenfunctions of L^+ and L^- defined by

$$L^{\pm} = \begin{pmatrix} b_1 & a_1 & & & \pm a_N \\ a_1 & b_2 & & & \\ & & \ddots & & \\ & & & b_{N-1} & a_{N-1} \\ \pm a_N & & & a_{N-1} & b_N \end{pmatrix}. \tag{4.1.28}$$

Since the elements of these symmetric matrices are real, the eigenvalues are also real. To prove this, let L be a symmetric matrix, and u its eigenfunction with the eigenvalue λ. If we denote by * the complex conjugate, then we have

$$\lambda u^* \cdot u = u^* \cdot L \cdot u$$
$$\lambda^* u \cdot u^* = u \cdot L \cdot u^* = u^* \cdot L \cdot u \, ,$$

and then

$$(\lambda - \lambda^*) u^* \cdot u = 0 \, .$$

Therefore λ is real. All the eigenvalues of (4.1.28) are thus real.

Equations (4.1.8, 9, 19) yield

$$\Delta(\lambda) = \left(\prod_{j=1}^{N} a_j \right)^{-1} \left[\lambda^N - \left(\sum_{j=1}^{N} b_j \right) \lambda^{N-1} + \cdots \right]. \tag{4.1.29}$$

Since $\Delta^2(\lambda) = 4$ has $2N$ roots $\lambda_l (l = 1, 2, \ldots, 2N)$, we have

$$\Delta^2(\lambda) - 4 = \left(\prod_{j=1}^{N} a_j \right)^{-2} \prod_{l=1}^{2N} (\lambda - \lambda_l) \, . \tag{4.1.30}$$

If λ_l's are numbered in increasing order, we see (cf. Fig. 4.2)

$$\lambda_1 < \lambda_2 \leq \lambda_3 < \lambda_4 \leq \lambda_5 < \cdots < \lambda_{2N-2} \leq \lambda_{2N-1} < \lambda_{2N} \, . \tag{4.1.31}$$

Only the intervals $[\lambda_{2l}, \lambda_{2l+1}]$ $(l = 1, 2, \ldots, N-1)$ may degenerate to one point yielding double roots. The spectrum (4.1.31) consists of two series interlaced, one coming from $\Delta - 2 = 0$ and the other from $\Delta + 2 = 0$ as shown below.

To prove the alternation of the spectral points, we use two solutions of (4.1.3a) with different λ,

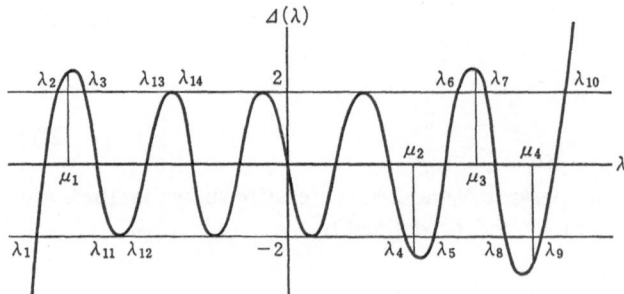

Fig. 4.2. Schematic diagram of $\Delta(\lambda) \sim \lambda$ (with motion) $n =$ odd $(= 11), g = 4$ (simple roots are numbered before double roots)

$$\varphi(n) = \varphi(n, \lambda). \qquad \psi(n) = \psi(n, \lambda') . \tag{4.1.32}$$

Then we see from (4.1.3a)

$$\begin{aligned}
(\lambda - \lambda')\varphi(n)\psi(n) &= \psi(n)[a_n\varphi(n + 1) + b_n\varphi(n) + a_{n-1}\varphi(n - 1)] \\
&\quad - \varphi(n)[a_n\psi(n + 1) + b_n\psi(n) + a_{n-1}\psi(n - 1)] \\
&= a_n[\varphi(n + 1)\psi(n) - \varphi(n)\psi(n + 1)] \\
&\quad - a_{n-1}[\varphi(n)\psi(n - 1) - \varphi(n - 1)\psi(n)] ,
\end{aligned} \tag{4.1.33}$$

so that

$$\begin{aligned}
(\lambda - \lambda') &\sum_{n=1}^{N} \varphi(n)\psi(n) \\
&= a_N[\varphi(N + 1)\psi(N) - \varphi(N)\psi(N + 1)] - a_0[\varphi(1)\psi(0) - \varphi(0)\psi(1)] .
\end{aligned} \tag{4.1.34}$$

Applying this to two fundamental solutions

$$\varphi(n) = \varphi_1(n, \lambda), \qquad \psi(n) = \varphi_1(n, \lambda') , \tag{4.1.35}$$

we have

$$\begin{aligned}
(\lambda - \lambda') &\sum_{n=1}^{N} \varphi_1(n, \lambda) \, \varphi_1(n, \lambda') \\
&= a_N[\varphi_1(N + 1, \lambda)\varphi_1(N, \lambda') - \varphi_1(N, \lambda)\varphi_1(N + 1, \lambda')] .
\end{aligned} \tag{4.1.36}$$

We let λ converge to λ', and write the derivative as

$$\varphi_1' = \frac{d\varphi_1}{d\lambda} , \tag{4.1.37}$$

then we have an expression for the norm

$$\|\varphi_1\|^2 = \sum_{n=1}^{N} \varphi_1^2(n)$$
$$= a_N[\varphi_1(N)\varphi_1'(N+1) - \varphi_1(N+1)\varphi_1'(N)]. \tag{4.1.38}$$

Similar expressions are obtained for φ_2, and for the scalar product $\varphi_1 \cdot \varphi_2$. If we solve these equations for $\varphi_1'(N+1)$, $\varphi_1'(N)$ etc., and substitute them into the derivative of $\Delta(\lambda)$ given by (4.1.19) with respect to λ, we have

$$\frac{d\Delta}{d\lambda} = \frac{1}{a_N} \sum_{n=1}^{N} \{\varphi_2(N)\varphi_1^2(n) - [\varphi_1(N) - \varphi_2(N+1)]\varphi_1(n)\varphi_2(n) - \varphi_1(N+1)\varphi_2^2(n)\}$$

$$\tag{4.1.39}$$

or, if we remember (4.1.15)

$$\frac{d\Delta}{d\lambda} = - \frac{\varphi_1(N+1)}{a_N} \sum_{n=1}^{N} \left\{ \left[\varphi_2(n) + \frac{\varphi_1(N) - \varphi_2(N+1)}{2\varphi_1(N+1)} \varphi_1(n) \right]^2 \right.$$
$$\left. - \frac{\Gamma}{4\varphi_1^2(N+1)} \varphi_1^2(n) \right\}, \tag{4.1.40}$$

where, by (4.1.15) and (4.1.19)

$$\Gamma = \varphi_1[(N) - \varphi_2(N+1)]^2 + 4\varphi_2(N)\varphi_1(N+1)$$
$$= [\varphi_1(N) + \varphi_2(N+1)]^2 - 4 \tag{4.1.41}$$
$$= \Delta^2 - 4.$$

Therefore, as long as $\Delta^2 - 4 < 0$, $d\Delta/d\lambda$ will have the sign of $-\varphi_1(N+1)$. If $\varphi_1(N+1)$ were to vanish when $\Delta^2 - 4 < 0$, (4.1.15) would indicate that $\varphi_1(N)\varphi_2(N+1) = 1$ and that $|\Delta| = |\varphi_1(N) + 1/\varphi_1(N)| \geq 2$, which would be a contradiction. Therefore $\varphi_1(N+1)$ cannot vanish as long as $\Delta^2 - 4 < 0$. In other words, $d\Delta/d\lambda$ can only change sign in the region where $\Delta^2 - 4 > 0$. This proves the alternation of the spectral points. Figure 4.2 illustrates a possible relationship between $\Delta(\lambda)$ and λ. As already mentioned, the roots of $\Delta^2 - 4 = 0$, or $\lambda_1, \lambda_2, \ldots, \lambda_{2N}$ are independent of time. Further it will be shown in Sect. 4.6 that the function $\Delta(\lambda)$ itself is independent of time.

Problem 4.1. Show that

$$\Delta(\lambda) = - \frac{1}{a_0} (\lambda - b_1)\varphi_1(N+1) + O(\lambda^{N-2}) = (-1)^N A^{-1} \det(L^{\pm} - \lambda I) \pm 2$$

where

$$A = a_1 a_2 \cdots a_N.$$

4.2 Auxiliary Spectrum

Though the total number of the roots λ_j (spectrum) of $\Delta^2(\lambda) - 4 = 0$ is $2N$, and is the same as the total number of dynamical variables a_n and b_n, we cannot determine these dynamical variables even if all the λ_j's are given. Indeed, it can be shown that λ_j's are not all independent (cf. Appendix C). Therefore to solve the inverse problem, we have to have further information. As such, we may use the auxiliary spectrum μ_j [4.2, 3], which is defined under boundary conditions different from those for λ_j.

We define the auxiliary spectrum μ_j by

$$\varphi_1(N + 1, \mu_j) = 0 \qquad (j = 1, 2, \cdots, N - 1). \tag{4.2.1}$$

Since φ_1 is a polynomial of order $N - 1$, there are $N - 1$ roots μ_j. According to the definition of φ_1, we have $\varphi_1(1, \mu_j) = 0$ [also $\varphi_1(0, \mu_j) = 1$, but this has no meaning in defining μ_j]. Since by (4.2.1) and (4.1.15)

$$\varphi_1(N, \mu_j)\varphi_2(N + 1, \mu_j) = 1 , \tag{4.2.2}$$

we have

$$\Delta(\mu_j) = \varphi_1(N, \mu_j) + \frac{1}{\varphi_1(N, \mu_j)} \tag{4.2.3}$$

by (4.1.9). Thus we see

$$|\Delta(\mu_j)| \geq 2 \tag{4.2.4}$$

which means that μ_j's lie in the unstable regions. All the μ_j's are simple (cf. Appendix D), and μ_j is between λ_{2j} and λ_{2j+1}:

$$\lambda_{2j} \leq \mu_j \leq \lambda_{2j+1} \qquad (j = 1, 2, ..., N-1). \tag{4.2.5}$$

Now, since by (4.1.8) and (4.2.1)

$$\varphi_1(N+1, \lambda) = - a_0 \left(\prod_{j=1}^N a_j\right)^{-1} \left[\lambda^{N-1} - \left(\sum_{j=2}^N b_j\right) \lambda^{N-2} + \cdots\right]$$

$$= - a_0 \left(\prod_{j=1}^N a_j\right)^{-1} \prod_{j=1}^{N-1} (\lambda - \mu_j) , \tag{4.2.6}$$

we obtain the relation

$$\sum_{j=2}^N b_j = \sum_{j=1}^{N-1} \mu_j \tag{4.2.7}$$

which may be rewritten as

$$\sum_{j=1}^{N} b_j - b_1 = \sum_{j=1}^{N-1} \mu_j \,, \tag{4.2.8}$$

On the other hand, comparing (4.1.29,30) we have

$$\bar{\Lambda} \equiv \sum_{j=1}^{N} b_j = \frac{1}{2} \sum_{j=1}^{2N} \lambda_j = \text{const. (indep. of time)} \,. \tag{4.2.9}$$

Thus, if all the μ_j's are known, b_1 is given as

$$b_1 = \bar{\Lambda} - \sum_{j=1}^{N-1} \mu_j \,, \tag{4.2.10}$$

When the curve $\Delta(\lambda)$ cuts $\Delta(\lambda) = \pm 2$ we have a simple root, and when it touches $\Delta(\lambda) = \pm 2$ we have a double root. When λ_{2j} and λ_{2j+1} coincide (double root), μ_j also coincides with them ($\lambda_{2j} = \mu_j = \lambda_{2j+1}$). When λ_{2j} and λ_{2j+1} differ simple roots), μ_j is between them (μ_j oscillates between λ_{2j} and λ_{2j+1} with time as will be shown later). In the following discussion, the simple roots of λ_j play a central role. Let their number be $2g + 2$, and changing the numbering we denote the simple roots by

$$\lambda_1 < \lambda_2 < \cdots < \lambda_{2g+2} \tag{4.2.11a}$$

and the double roots by

$$\lambda_{2j+1} = \lambda_{2j+2} \quad (j = g + 1, \cdots, N - 1) \,. \tag{4.2.11b}$$

We also change the numbering of μ_j so that

$$\lambda_{2j} < \mu_j < \lambda_{2j+1} \quad (j = 1, \cdots, g) \tag{4.2.12a}$$

and, of course, for the remaining double roots

$$\mu_j = \lambda_{2j+1} = \lambda_{2j+2} \quad (j = g + 1, \cdots, N - 1) \,. \tag{4.2.12b}$$

Using these definitions, (4.2.10) is rewritten as

$$b_1 = \Lambda - \sum_{j=1}^{g} \mu_j \tag{4.2.13}$$

with

$$\Lambda = \frac{1}{2} \sum_{j=1}^{2g+2} \lambda_j = \text{const} \,. \tag{4.2.14}$$

Therefore, if we know the auxiliary spectrum $\mu_j \, (j = 1, 2, \cdots, g)$, b_1 is obtained.

A similar argument will lead to a formula for b_k if we shift all the suffixes n by a constant k [4.3]. Thus let $\varphi_1(n|k) = \varphi_1(n, \lambda|k)$ and $\varphi_2(n|k) = \varphi_2(n, \lambda|k)$ denote the solution of

$$a_{n+k-1}\varphi(n-1|k) + b_{n+k}\varphi(n|k) + a_{n+k}\varphi(n+1|k) = \lambda\varphi(n|k) \qquad (4.2.15)$$

subject to the boundary conditions

$$\left.\begin{array}{ll} \varphi_1(0|k) = 1, & \varphi_1(1|k) = 0 \\ \varphi_2(0|k) = 0, & \varphi_2(1|k) = 1 \end{array}\right\} \qquad (4.2.16)$$

In terms of $\varphi_1(n)$ and $\varphi_2(n)$, we can express $\varphi_1(n|k)$ and $\varphi_2(n|k)$ as

$$\left.\begin{array}{l} \varphi_1(n|k) = a_1\varphi_1(k+n) + \beta_1\varphi_2(k+n) \\ \varphi_2(n|k) = \alpha_2\varphi_1(k+n) + \beta_2\varphi_2(k+n) \end{array}\right\} \qquad (4.2.17)$$

For $n = 0$ and $n = 1$, we have

$$\left.\begin{array}{l} 1 = \alpha_1\varphi_1(k) + \beta_1\varphi_2(k) \\ 0 = \alpha_1\varphi_1(k+1) + \beta_1\varphi_2(k+1) \\ 0 = \alpha_2\varphi_1(k) + \beta_2\varphi_2(k) \\ 1 = \alpha_2\varphi_1(k+1) + \beta_2\varphi_2(k+1) \end{array}\right\}. \qquad (4.2.18)$$

Therefore, eliminating β_1 from the first two equations we have

$$\begin{aligned} \varphi_2(k+1) &= \alpha_1[\varphi_1(k)\varphi_2(k+1) - \varphi_1(k+1)\varphi_2(k)] \\ &= \frac{W}{a_k}\alpha_1 \\ &= \frac{a_0}{a_k}\alpha_1 \end{aligned} \qquad (4.2.19a)$$

where we have used the Wronskian W (4.1.14). Similarly

$$\left.\begin{array}{l} \varphi_1(k+1) = -\dfrac{W}{a_k}\beta_1 = -\dfrac{a_0}{a_k}\beta_1 \\[2mm] \varphi_2(k) = -\dfrac{W}{a_k}\alpha_2 = -\dfrac{a_0}{a_k}\alpha_2 \\[2mm] \varphi_1(k) = \dfrac{W}{a_k}\beta_2 = \dfrac{a_0}{a_k}\beta_2 \end{array}\right\}. \qquad (4.2.19b)$$

Thus

$$\left.\begin{array}{l} \varphi_1(n|k) = \dfrac{a_k}{a_0}[\varphi_2(k+1)\varphi_1(k+n) - \varphi_1(k+1)\varphi_2(k+n)] \\[2mm] \varphi_2(n|k) = \dfrac{a_k}{a_0}[-\varphi_2(k)\varphi_1(k+n) + \varphi_1(k)\varphi_2(k+n)] \end{array}\right\}, \qquad (4.2.20)$$

It is easy to show that the discriminant is invariant

$$\Delta(\lambda|k) = \varphi_1(N|k) + \varphi_2(N+1|k)$$
$$= \Delta(\lambda),$$
(4.2.21)

which means that the roots of $\Delta^2 - 4 = 0$ are invariant

$$\lambda_j(k) = \lambda_j(0) = \lambda_j.$$
(4.2.22)

Then, we define $\mu_j(k)$ by

$$\varphi_1[N+1, \mu_j(k)|k] = 0.$$
(4.2.23)

By an argument similar to the above, we see that

$$\lambda_{2j} \le \mu_j(k) \le \lambda_{2j+1},$$
(4.2.24)

but in general

$$\mu_j(k) \ne \mu_j[= \mu_j(0)].$$
(4.2.25)

Similarly to (4.1.8), we have

$$\varphi_1(N+1, \lambda|k) = -a_k\left(\prod_{j=1}^{N} a_{j+k}\right)^{-1}\left[\lambda^{N-1} - \left(\sum_{j=2}^{N} b_{j+k}\right)\lambda^{N-2} + \cdots\right]$$
$$= -a_k\left(\prod_{j=1}^{N} a_{j+k}\right)^{-1}\prod_{l=1}^{N-1}[\lambda - \mu_l(k)].$$
(4.2.26)

Therefore, by virtue of

$$\sum_{j=2}^{N} b_{j+k} = b_{k+3} + b_{k+2} + \cdots + b_N + b_{N+1} + \cdots + b_{k+N-1} + b_{k+N}$$
$$= b_{k+2} + b_{k+3} + \cdots + b_N + b_1 + \cdots + b_{k-1} + b_k$$
$$= \sum_{l=1}^{N} b_l - b_{k+1}$$
(4.2.27)

we have

$$\sum_{l=1}^{N} b_l - b_{k+1} = \sum_{j=1}^{N-1} \mu_j(k).$$
(4.2.28)

On the other hand, we have (4.2.9), or

$$\bar{\Lambda} = \frac{1}{2}\sum_{j=1}^{2N} \lambda_j = \sum_{j=1}^{N} b_j.$$
(4.2.29)

Thus we are led to the formula

$$b_{k+1} = \tilde{\Lambda} - \sum_{j=1}^{N-1} \mu_j(k) \,. \tag{4.2.30}$$

We change the numbering so that $\mu_j(k)$ $(j = 1, 2, \ldots, g)$ is between simple roots; then

$$\left.\begin{array}{ll} \lambda_{2j} \leq \mu_j(k) \leq \lambda_{2j+1} & (j = 1,2, \cdots, g) \\ \mu\,(k) = \lambda_{2j+1} = \lambda_{2j+2} & (j = g + 1, g + 2, \cdots, N - 1) \end{array}\right\} \,. \tag{4.2.31}$$

Using these definitions, we have the important formula

$$b_{k+1} = \Lambda - \sum_{j=1}^{g} \mu_j(k) \tag{4.2.32a}$$

with

$$\Lambda = \frac{1}{2} \sum_{j=1}^{2g+2} \lambda_j \,. \tag{4.2.32b}$$

4.3 Relation Between $\mu_j(k)$ and $\mu_j(0)$

While the roots λ_j of $\Delta^2(\lambda) - 4 = 0$ are independent of time, each of $\mu_j(k)$ will oscillate in the interval $[\lambda_{2j+1}, \lambda_{2j+2}]$ $(j = 1,2, \ldots, g)$ (cf. Sect. 4.6). Before solving the motion, we shall note an important relation between $\mu_j(k)$ and $\mu_j(0)$, and proceed the clarify relations between the dynamical quantities (a_n, b_n) and $\mu_j(k)$ in Sects. 4.4, 6.

The Bloch function defined by (4.1.16) may be written, except for a constant factor, as

$$\varphi^{\pm}(n) = \frac{c_1}{c_2} \varphi_1(n) + \varphi_2(n) \tag{4.3.1}$$

where c_1/c_2 is given by solving (4.1.17) as

$$\frac{c_1}{c_2} = \frac{p - \varphi_2(N + 1)}{\varphi_1(N + 1)} = \frac{\varphi_1(N) - \varphi_2(N + 1) \pm \sqrt{\Delta^2 - 4}}{2\varphi_1(N + 1)} \,, \tag{4.3.2}$$

Alternatively we may write

$$\frac{c_1}{c_2} = \frac{\varphi_2(N)}{p - \varphi_1(N)} = \frac{2\varphi_2(N)}{-\varphi_1(N) + \varphi_2(N + 1) \pm \sqrt{\Delta^2 - 4}} \,. \tag{4.3.3}$$

Equating these two expressions we obtain(4.1.41) once again. On the other hand using (4.3.1,2) and (4.1.41) we have

$$\varphi^+(n)\varphi^-(n)$$

$$= \frac{1}{4\varphi_1^2(N+1)} \{[\varphi_1(N) - \varphi_2(N+1) + \sqrt{\Delta^2 - 4}]\,\varphi_1(n) + 2\varphi_1(N+1)\varphi_2(n)$$

$$\cdot \{[\,\{\varphi_1(N) - \varphi_2(N+1) - \sqrt{\Delta^2 - 4}\}]\,\varphi_1(n) + 2\varphi_1(N+1)\varphi_2(n)\}$$

$$= \frac{1}{\varphi_1(N+1)} \{[\varphi_1(N) - \varphi_2(N+1)]\,\varphi_1(n)\varphi_2(n) + \varphi_1(N+1)\varphi_2^2(n) - \varphi_2(N)\varphi_1^2(n)\}$$

$$= \frac{a_0/a_{n-1}}{\varphi_1(N+1)}\,\varphi_1(N+1\,|\,n-1) \tag{4.3.4}$$

where, for the last line, we have used (4.2.20) and (4.1.10, 11). Changing n to $n+1$, we have

$$\varphi^+(n+1)\varphi^-(n+1) = \frac{a_0}{a_n}\frac{\varphi_1(N+1\,|\,n)}{\varphi_1(N+1)}. \tag{4.3.5}$$

Now, by the definition of $\mu_j = \mu_j(0)$, we have $\varphi_1(N+1) = 0$ at $\lambda = \mu_j$, where, by (4.1.41), the numerator of (4.3.2) for the $+$ sign is

$$\varphi_1(N, \mu_j) - \varphi_2(N+1, \mu_j) + \sqrt{\Delta^2(\mu_j) - 4} = 2\sqrt{\Delta^2(\mu_j) - 4} \tag{4.3.6}$$

which does not vanish in general. Therefore $\varphi^+(n+1)$ has a pole at $\lambda = \mu_j$. On the other hand, $\varphi_1(N+1\,|\,n)$ vanishes at $\lambda = \mu_j(n)$, so that by (4.3.5), we see that $\varphi^-(n+1)$ has a zero at $\lambda = \mu_j(n)$. Thus writing k for n, we have

$$\varphi^+(k+1, \mu_j) = \infty, \qquad \varphi^-[k+1, \mu_j(k)] = 0. \tag{4.3.7}$$

We can consider $\varphi^+(k+1, \lambda)$ and $\varphi^-(k+1, \lambda)$ as values of the function $\varphi(k+1, \lambda)$ on the two sheets of the Riemann surface S with branch cuts along the intervals between λ_{2j-1} and $\lambda_{2j}(j = 1, 2, \ldots, g)$, the zeros of $[\Delta^2(\lambda) - 4]^{1/2}$. On the upper sheet $[\Delta^2(\lambda) - 4]^{1/2}$ has the value $\sqrt{\Delta^2(\lambda) - 4}$ and on the lower sheet $-\sqrt{\Delta^2(\lambda) - 4}$. Thus the Bloch function $\varphi(k+1, \lambda)$ has simple zeros at $\mu_j(k)$ and simple poles at $\mu_j(0)$ on S, and connects these as we expected.

Further, $\varphi(k+1, \lambda)$ has a zero and a pole at infinity on the Riemann surface. In fact, for sufficiently large λ, since $\varphi_1(N) \sim \lambda^{N-2}$, $\varphi_2(N) \sim \lambda^{N-1}$, $\varphi_1(N+1) \sim \lambda^{N-1}$, $\varphi_2(N+1) \sim \lambda^N$, and $\Delta(\lambda) = \varphi_1(N) + \varphi_2(N+1) \sim \lambda^N$,

$$\varphi^+(k+1) \sim \varphi_2(k+1) \sim \lambda^k \tag{4.3.8}$$

and by (4.3.5,8) and (4.2.26) we have

$$\varphi^-(k+1) \sim \frac{\varphi_1(N+1\,|\,k)}{\varphi_1(N+1)\varphi^+(k+1)} \sim \lambda^{-k} . \tag{4.3.9}$$

Therefore $\varphi(k+1)$ has a pole of the kth order at ∞(on the upper sheet), and a zero of the kth order at ∞'(infinity on the lower sheet).

Consider the differential

$$\omega(k) = \left[\frac{d}{d\lambda} \ln \varphi(k+1, \lambda) \right] d\lambda \tag{4.3.10}$$

which has poles at $\mu_j(0)$ and $\mu_j(k)$ with residues $+1$ and -1, respectively, and poles at ∞ and ∞' with residues $+k$ and $-k$, respectively, [4.3]. In fact, using (4.3.1,2) we see that

$$\omega(k) = \left[f(\lambda) + g(\lambda)(\Delta^2 - 4)^{1/2} + \frac{h(\lambda)}{(\Delta^2 - 4)^{1/2}} \right] \frac{d\lambda}{F(\lambda)} \tag{4.3.11}$$

where $f(\lambda)$, $g(\lambda)$, $h(\lambda)$, and $F(\lambda)$ are certain polynomials. But their expressions are not necessary here.

Since there are $2g+2$ simple roots among $2N$ roots of $\Delta^2(\lambda)-4=0$, we may write

$$[\Delta^2(\lambda) - 4]^{1/2} = \text{(polynomial of } \lambda) \times [R(\lambda)]^{1/2} \tag{4.3.12a}$$

with

$$R(\lambda) = \prod_{j=1}^{2g+2} (\lambda - \lambda_j) . \tag{4.3.12b}$$

We introduce the differential

$$\omega_{(s)} = \frac{\lambda^s d\lambda}{[R(\lambda)]^{1/2}} \tag{4.3.13}$$

and the base $\{\omega_l\}$ of hyperelliptic functions

$$\omega_l = \sum_{s=0}^{g-1} c_{ls} \omega_{(s)} . \tag{4.3.14}$$

Such differentials have no pole, and are called Abelian differentials of the first kind (a differential with poles but vanishing residue is called an Abelian differential of the second kind, and a differential with nonvanishing residue is called an Abelian differential of the third kind).

The base $\{\omega_l\}$ is called the normalized differential of the first kind when the coefficients c_{ls} are normalized in such a way that

$$\int_{\alpha_j} \omega_l = \delta_{jl} \qquad (ji, l = 1,2, \cdots, g) \,. \tag{4.3.15}$$

Further we put

$$\int_{\beta_j} \omega_l = \tau_{jl} \qquad (j, l = 1,2, \cdots, g) \,. \tag{4.3.16}$$

Here α_j is a closed contour which surrounds the cut $(\lambda_{2j+1}, \lambda_{2j+2})$ on the upper sheet, while β_j is a closed contour which starts at λ_2, goes on the lower sheet as far as λ_{2j+1}, crosses to the upper sheet, and ends at λ_2 (Fig. 4.3). $\int_\alpha \omega$ is called an α-period, while $\int_\beta \omega$ is called a β-period. $[R(\lambda)]^{1/2}$ is real at $\lambda \to \infty$ on the real axis. Thus, $[R(\lambda)]^{1/2}$ is pure imaginary between λ_{2j+1} and λ_{2j+2} on the real axis, and therefore the coefficients c_{ls} are pure imaginary. Consequently τ_{jk} are also pure imaginary [we have to note that $\sqrt{-1}\,\delta_{jl}$ is used in certain books for the right-hand side of (4.3.15)]. It can be shown that $\tau_{jk} = \tau_{kj}$ (cf. Problem in the next Chapter.).

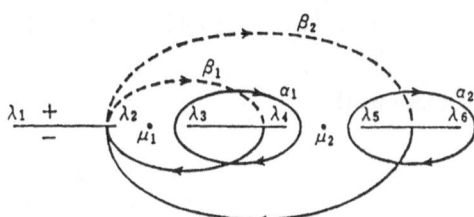

Fig. 4.3. α-periods and β-periods ($g = 2$)

By the residue theorem, the sum of all the residues vanishes. Let the Abelian differential of the third kind, with the residue 1 and -1 at P and Q respectively, be denoted by $\omega(P, Q)$. We can add certain differentials of the first kind to the differential of the third kind so that all the α-periods vanish,

$$\int_{\alpha_j} \omega(P, Q) = 0 \,. \tag{4.3.17a}$$

Then $\omega(P, Q)$ is called the normal differential of the third kind.

The differential (4.3.10), or (4.3.11), can be expressed as a linear combination of the differential Ω of the second kind, the normal differential $\omega(P, Q)$ of the third kind and the normalized differential ω_j of the first kind. Noting the poles, we can write (4.3.10) as

$$\omega(k) = \Omega + k\omega(\infty', \infty) + \sum_{j=1}^{g} \omega[\mu_j(k), \mu_j(0)] + \sum_{j=1}^{g} c_j \omega_j \,, \tag{4.3.18}$$

c_j being some complex numbers.

Since in (4.3.10), where $\omega(k)$ is defined, $\varphi(k+1, \lambda)$ is a single valued function of λ on the Riemann surface S, we have

$$\int\limits_{\alpha_l} \omega(k) = 2\pi n_l i \tag{4.3.19}$$

$$\int\limits_{\beta_l} \omega(k) = 2\pi m_l i \tag{4.3.20}$$

where n_l and m_l are certain integers. And since $\omega[\mu_j(k), \mu_j(0)]$ is normalized so that

$$\int\limits_{\alpha_l} \omega[\mu_j(k), \mu_j(0)] = 0 , \tag{4.3.17b}$$

(4.3.18) yields in virtue of (4.3.15)

$$c_j = 2\pi n_j i . \tag{4.3.21}$$

Therefore, the β_l-integral of (4.3.18) gives

$$k \int\limits_{\beta_l} \omega(\infty', \infty) + \sum_{j=1}^{g} \int\limits_{\beta_l} \omega(\mu_j(k), \mu_j(0)] + 2\pi i \sum_{j=1}^{g} n_j \int\limits_{\beta_l} \omega_j = 2\pi m_l i . \tag{4.3.22}$$

However, it can be shown (see the next section) that

$$\int\limits_{\beta_l} \omega[\mu_j(k), \mu_j(0)] = 2\pi i \int\limits_{\mu_j(0)}^{\mu_j(k)} \omega_l \tag{4.3.23}$$

where ω_l is the normalized differential of the first kind defined by (4.3.14). Thus (4.3.22) yields

$$k \int\limits_{\infty}^{\infty'} \omega_l + \sum_{j=1}^{g} \int\limits_{\mu_j(0)}^{\mu_j(k)} \omega_l = - \sum_{j=1}^{g} n_j \tau_{lj} + m_l , \tag{4.3.24}$$

Though the first term on the left-hand side vanishes for $k=0$, the second term depends on the path of integration and is equal, for $k=0$, to the right-hand side, which does not depend on k. We have therefore

$$\sum_{j=1}^{g} \int\limits_{\mu_0}^{\mu_j(k)} \omega_l = k \int\limits_{\infty'}^{\infty} \omega_l + \sum_{j=1}^{g} \int\limits_{\mu_0}^{\mu_j(0)} \omega_l - \sum_{j=1}^{g} n_j \tau_{lj} + m_l \qquad (l = 1, 2, \cdots, g) \tag{4.3.25}$$

where μ_0 is a fixed point on S which can be chosen arbitrarily. This is the required relationship between $\mu_j(k)$ and $\mu_j(0)$. If λ_j's are given and if we know $\mu_j(0)$, the right-hand side of (4.3.25) is a known quantity. To indicate that this set of equations actually determines g unknowns, $\mu_j(k)$ is called the Jacobi inversion problem. In our case it is not necessary to obtain each $\mu_j(k)$, but we have to find $\sum_{j=1}^{g} \mu_j(k)$ in (4.3.32) to express b_k as a function of k.

4.4 Related Integrals on the Riemann Surface

Our next task is to solve the Jacobi inversion problem, and since it is related to Abelian differentials, we shall investigate them a little further in this section providing some theorems [4.4]. In what follows, we shall frequently make use of the Cauchy integral theorem on a simply connected region S_0 associated with the Riemann surface S.

Our Riemann surface consists of two sheets of complex planes joined along the branch cuts $[\lambda_1, \lambda_2]$, $[\lambda_2, \lambda_3]$, ... We first make two complex λ spheres by stereographic projection of each sheet (Fig. 4.4). Along the banks of the cuts we put $+$ and $-$ signs: the $+$ signs refer to the positive side of the imaginary axis, and the $-$ signs to the negative side.

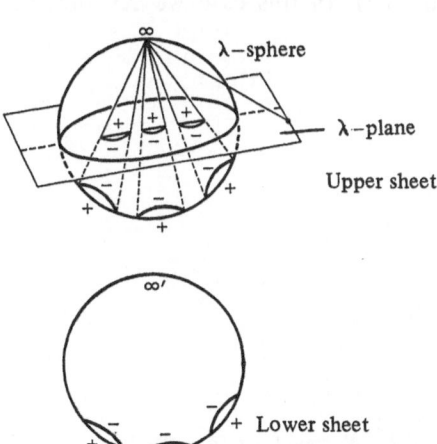

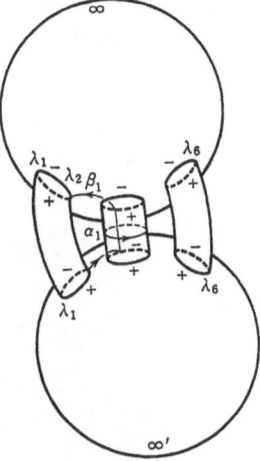

Fig. 4.4. Stereographic projection of two Riemann sheets

Fig. 4.5. Topological mapping of the Riemann surface

We turn over one of the λ spheres, until the corresponding branch cuts of the two spheres face each other, the $+(-)$ banks facing the $-(+)$ banks of the other sphere. We open the cuts widely and join the facing banks by $g+1$ tubes and paste (Fig. 4.5).

By topological deformation we have a surface consisting of handles (tubes) and a sphere which is made from the two facing spheres and the tube for the branch cut $[\lambda_1, \lambda_2]$ (see Fig. 4.6a).

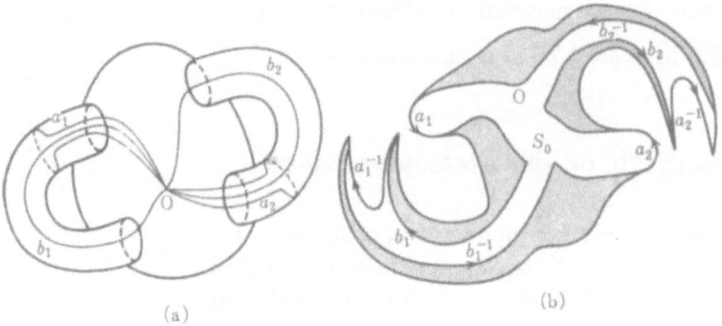

(a)   (b)

Fig. 4.6. Canonical dissection ($g = 2$)

Then, by using scissors, we cut and open along the lines a_1, b_1, a_2, b_2, ... in Fig. 4.6a through a point 0. Thus we obtain a simply connected region S_0 as shown in Fig. 4.6b. We specify the edges of this region by the arrows a_1, b_1, a_1^{-1}, b_1^{-1}, a_2, b_2, ... a_g^{-1}, b_g^{-1}, and flatten the surface to the normal form $a_1 b_1 a_1^{-1} ... a_g^{-1} b_g^{-1}$ of the surface S_0 (Fig. 4.7). In this case we say that our Riemann surface is of genus g.

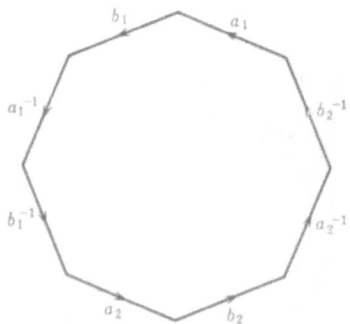

Fig. 4.7. Normal form ($g = 2$)

a_1 gives the path of integration α_1, and b_1 the path β_1 and so forth. a_1^{-1} gives the path of integration α_1^{-1} which is the reverse of α_1, and b_1^{-1} the reverse of β_1, and so forth.

I) Let ω and η be differentials with the "periods" (A_i, B_i) and (A_i', B_i') $(i+1, 2, ... , g)$,

$$\left. \begin{array}{ll} \int_{\alpha_i} \omega = A_i & \int_{\beta_i} \omega = B_i \\ \int_{\alpha_i} \eta = A_i' & \int_{\beta_i} \eta = B_i' \end{array} \right\} , \qquad (4.4.1)$$

Let Q' be a point on the line a_i^{-1} which corresponds to Q on a_i (Fig. 4.8a), and let

$$\omega = df.\tag{4.4.2}$$

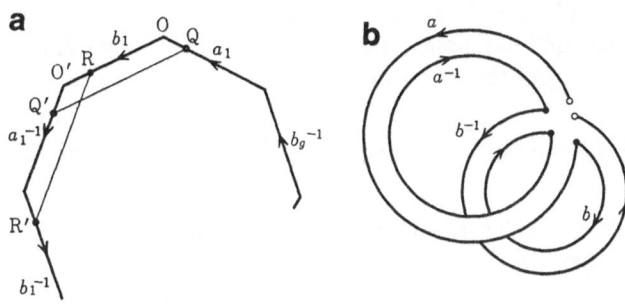

Fig. 4.8.
Paths of integration

Then we have

$$
\begin{aligned}
\int_{QQ'} \omega &= (\int_{QO} + \int_{OO'} + \int_{O'Q'}) \, df \\
&= \int_{OO'} df = \int_{\beta_i} df = B_i \\
B_i &= f(Q') - f(Q)
\end{aligned}
\right\}\tag{4.4.3}
$$

for

$$\int_{QO} df = - \int_{O'Q'} df.$$

Similarly,

$$-A_i = f(R') - f(R)\tag{4.4.4}$$

where R' on b_i^{-1} is the point corresponding to R on b_i.

Now, since $f(Q') = f(Q) + B_i$ and as differentials

$$\eta(Q') = -\eta(Q)\tag{4.4.5}$$

and so on, we have

$$
\begin{aligned}
\int_{a_i+\beta_i+a_i^{-1}+\beta_i^{-1}} f\eta &= \int_{a_i} f\eta + \int_{\beta_i} f\eta - \int_{a_i}(f+B_i)\,\eta - \int_{\beta_i}(f-A_i)\eta \\
&= -B_i \int_{a_i} \eta + A_i \int_{\beta_i} \eta \\
&= A_i B'_i - B_i A'_i.
\end{aligned}\tag{4.4.6}
$$

Therefore

$$\int_C f\eta = \sum_{j=1}^{g} (A_j B'_j - B_j A'_j) \tag{4.4.7}$$

where the contour C encircles the polygon $a_1 b_1 \ldots a_g^{-1} b_g^{-1}$.

Let $\omega = df = \omega^{[1]}$ be a differential of the first kind and $\eta = \omega^{[3]}$ be of the third kind, and assume that $\omega^{[3]}$ has poles at P_l with residues C_l ($l=1,2,\ldots,m$). Then, by the Cauchy integral theorem, we have

$$\sum_{j=1}^{g} (A_j B'_j - B_j A'_j) = \int_C f\omega^{[3]} = 2\pi i \sum_{l=1}^{m} f(P_l) C_l = 2\pi i \sum_{l=1}^{m} C_l \int_{P_0}^{P_l} \omega^{[1]} \tag{4.4.8}$$

where P_0 is a point on S_0, and $f(P_l) = \int_{P_0}^{P_l} df = \int_{P_0}^{P_l} \omega^{[1]}$. Further, let $\omega^{[3]}$ be the normal differential so that

$$A'_j = 0 \text{ (for all } j) \tag{4.4.9}$$

and $\omega^{[1]} = \omega_k$ be the normalized differential of the first kind (4.3.15) such that

$$A_j = \int_{\alpha_j} \omega_k = \delta_{jk} . \tag{4.4.10}$$

Then, by (4.4.8),

$$B'_k = 2\pi i \sum_{l=1}^{m} C_l \int_{P_0}^{P_l} \omega_k \tag{4 4.11}$$

or

$$\int_{\beta_k} \omega^{[3]} = 2\pi i \sum_{l=1}^{m} C_l \int_{P_0}^{P_l} \omega_k . \tag{4.4.12}$$

If $\omega^{[3]}$ has poles at $P[=\mu_j(k)]$ with residue $C_p = 1$ and $Q[=\mu_j(0)]$ with residue $C_Q = -1$, then (4.4.12) reduces to

$$\int_{\beta_k} \omega^{[3]} = 2\pi i \left(\int_{P_0}^{P} \omega_k - \int_{P_0}^{Q} \omega_k \right) \tag{4.4.13}$$

or

$$\int_{\beta_k} \omega^{[3]} = 2\pi i \int_{Q}^{P} \omega_k \tag{4.4.14}$$

which proves (4.3.23).

II) We define the multidimensional ϑ function (the Riemann ϑ function) by

$$\vartheta(u) = \sum_{m_1, \cdots, m_g = -\infty}^{\infty} \exp\left(2\pi i \sum_{j=1}^{g} m_j u_j + \pi i \sum_{j,k=1}^{g} \tau_{jk} m_j m_k\right). \tag{4.4.15a}$$

This can be written as

$$\vartheta(u) = \sum_{m} \exp\left(2\pi i m \cdot u + \pi i m \cdot \tau \cdot m\right) \tag{4.4.15b}$$

where n and m are vectors and τ is a matrix

$$u = \begin{pmatrix} u_1 \\ \vdots \\ u_g \end{pmatrix}, \quad m = \begin{pmatrix} m_1 \\ \vdots \\ m_g \end{pmatrix}, \quad \tau = \begin{pmatrix} \tau_{11} & \tau_{12} & \cdots & \tau_{1g} \\ \vdots & & & \\ \tau_{g1} & \tau_{g2} & \cdots & \tau_{gg} \end{pmatrix}. \tag{4.4.16}$$

It is convenient to introduce the notations

$$u + e_k = \begin{pmatrix} u_1 \\ \vdots \\ u_k + 1 \\ u_{k+1} \\ \vdots \\ u_g \end{pmatrix} \tag{4.4.17a}$$

$$u + \tau_k = \begin{pmatrix} u_1 + \tau_{1k} \\ \vdots \\ u_k + \tau_{kk} \\ u_g + \tau_{gk} \end{pmatrix}. \tag{4.4.17b}$$

We easily see that

$$\left.\begin{aligned} \vartheta(u + e_k) &= \vartheta(u) \\ \vartheta(u + \tau_k) &= e^{-2\pi i u_k - \pi i \tau_{kk}} \vartheta(u) \end{aligned}\right\}. \tag{4.4.18}$$

The last equality is derived by replacing m_k by $m_k + 1$. If we introduce

$$F(u) = \vartheta(u - \bar{c}) \tag{4.4.19a}$$

where $\bar{c}$ is a constant vector

$$\bar{c} = \begin{pmatrix} \bar{c}_1 \\ \vdots \\ \bar{c}_g \end{pmatrix}, \tag{4.4.19b}$$

we have

$$\left.\begin{aligned} F(u + e_k) &= F(u) \\ F(u + \tau_k) &= e^{-2\pi i (u_k - c_k) - \pi i \tau_{kk}} F(u) \end{aligned}\right\}, \tag{4.4.20}$$

Now, let

$$u_l(P) = \int_{P_0}^{P} \omega_l \tag{4.4.21a}$$

or

$$du_l = \omega_l = \sum_{s=0}^{g-1} c_{ls} \frac{\lambda^s d\lambda}{[R(\lambda)]^{1/2}}, \tag{4.4.21b}$$

where P is a point on the Riemann surface. Further we write

$$f(P) = \vartheta[\boldsymbol{u}(P) - \bar{c}] \tag{4.4.22}$$

for F in this case, and let P_j be the zeros of $f(P)$,

$$f(P_j) = 0. \tag{4.4.23}$$

III) In the vicinity of P_j, we have $f \rightarrow r \exp(i\theta)$, and $\int df/f = i \int d\theta$. Therefore, for the contour C along the edge of S_0, the integral

$$n(f) = \frac{1}{2\pi i} \int_C \frac{df}{f} \tag{4.4.24}$$

gives the number of zeros P_j.

Let u_j^+ and u_j^- be the values at corresponding points on a_k and a_k^{-1}, or b_k and b_k^{-1}. If P is on a_k (cf. Fig. 4.8)

$$u_j^- = u_j^+ + \int_{\beta_k} \omega_j = u_j^+ + \tau_{kj} \tag{4.4.25}$$

and

$$f^- = f(u^+ + \tau_k) = e^{-2\pi i(u_k - c_k) - \pi i \tau_{kk}} f^+ \tag{4.4.26}$$

$$df^- = e^{-2\pi i(u_k - c_k) - \pi i \tau_{kk}}(df^+ - 2\pi i f^+ du_k). \tag{4.4.27}$$

Similarly, if P is on b_k

$$u_j^- = u_j^+ + \int_{\alpha_k^-} \omega_j = u_j^+ - \delta_{jk} \tag{4.4.28}$$

and

$$f^- = f(u^+ - e_k) = f^+. \tag{4.4.29}$$

Then we have

on a_k $\dfrac{df^-}{f^-} = \dfrac{df^+}{f^+} - 2\pi i \omega_k$

on b_k $\dfrac{df^-}{f^-} = \dfrac{df^+}{f^+}$

(4.4.30)

Therefore, (4.4.24) gives

$$n(f) = \frac{1}{2\pi i} \sum_{k=1}^{g} \left(\int_{\alpha_k} + \int_{\beta_k} \right) \left(\frac{df^+}{f^+} - \frac{df^-}{f^-} \right)$$

$$= \frac{1}{2\pi i} \sum_{k=1}^{g} 2\pi i \int_{\alpha_k} \omega_k$$

$$= \sum_{k=1}^{g} \delta_{kk} = g \ .$$

(4.4.31)

Thus, we see that the number of zeros P_j of $f(P)$ is equal to the genus g.

IV) Since P_j's are the zeros of $f(P)$, we may write

$$\sum_{j=1}^{g} u_l(P_j) = \frac{1}{2\pi i} \int_C u_l \frac{df}{f} \ .$$

(4.4.32)

Making use of (4.4.25, 26, 30) we rewrite the integral on the right-hand side as

$$\int_C u_l \frac{df}{f} = \sum_{k=1}^{g} \left(\int_{\alpha_k} + \int_{\beta_k} \right) \left(u_l^+ \frac{df^+}{f^+} - u_l^- \frac{df^-}{f^-} \right)$$

$$= \sum_{k=1}^{g} \int_{\alpha_k} \left[u_l^+ \frac{df^+}{f^+} - (u_l^+ + \tau_{lk}) \left(\frac{df^+}{f^+} - 2\pi i \omega_k \right) \right]$$

$$+ \sum_{k=1}^{g} \int_{\beta_k} \left[u_l^+ \frac{df^+}{f^+} - (u_l^+ - \delta_{lk}) \frac{df^+}{f^+} \right]$$

$$= 2\pi i \sum_{k=1}^{g} \tau_{lk} \int_{\alpha_k} \omega_k - \sum_{k=1}^{g} \tau_{lk} \int_{\alpha_k} \frac{df^+}{f^+}$$

$$+ 2\pi i \sum_{k=1}^{g} \int_{\alpha_k} u_l^+ \omega_k + \int_{\beta_l} \frac{df^+}{f^+} \ .$$

(4.4.33)

On α_k, if we start from $Q_0^{(k)}$ and end at $Q_1^{(k)}$, then

$$f^+[u(Q_1^{(k)})] = f^+[u(Q_0^{(k)}) + e_k] = f^+[u(Q_0^{(k)})]$$

(4.4.34)

by (4.4.20). Therefore

$$\int_{\alpha_k} \frac{df^+}{f^+} = \ln f^+(Q_1^{(k)}) - \ln f^+(Q_0^{(k)}) = 0 \ .$$

(4.4.35)

Similarly on β_l, if we start from $\bar{Q}_0^{(k)}$ and end at $\bar{Q}_1^{(k)}$,

$$f^+[u(\bar{Q}_1^{(l)})] = f^+[u(\bar{Q}_0^{(l)}) + \tau_l]$$
$$= \exp\{-2\pi i[u(\bar{Q}_0^{(l)}) - \bar{c}_l] - \pi i \tau_{ll}\} f^+[u(\bar{Q}_0^{(l)})] \tag{4.4.36}$$

so that

$$\int_{\beta_l} \frac{df^+}{f^+} = \ln f^+[u(\bar{Q}_1^{(l)})] - \ln f^+[u(\bar{Q}_0^{(l)})]$$
$$= -2\pi i[u_l(\bar{Q}_0^{(l)}) - \bar{c}_l] - \pi i \tau_{ll} . \tag{4.4.37}$$

Thus, using (4.4.32, 33), we have a very important formula

$$\sum_{j=1}^{g} u_l(P_j) = \bar{c}_l - K_l + \sum_{k=1}^{g} \tau_{lk} \tag{4.4.38a}$$

where K_l is a constant given by

$$K_l = -\sum_{k=1}^{g} \int_{\alpha_k} u_l^+ \omega_k + \frac{1}{2}\tau_{ll} + u_l(\bar{Q}_0^{(l)}) . \tag{4.4.38b}$$

V) We consider the integral $\int \lambda \, df/f$. By (4.4.30)

$$\frac{1}{2\pi i} \int_C \lambda \frac{df}{f} = \frac{1}{2\pi i} \sum_{k=1}^{g} \left(\int_{\alpha_k} + \int_{\beta_k} \right) \lambda \left(\frac{df^+}{f^+} - \frac{df^-}{f^-} \right) = \sum_{k=1}^{g} \int_{\alpha_k} \lambda \omega_k . \tag{4.4.39}$$

On the other hand, by residue calculation we have

$$\frac{1}{2\pi i} \int_C \lambda \frac{df}{f} = \sum_{j=1}^{g} \lambda(P_j) + \text{Res}\{\infty\} + \text{Res}\{\infty'\} \tag{4.4.40}$$

where $\lambda(P_j)$ is the value of λ at the zero P_j of $f(P) = \vartheta[u(P - \bar{c})]$. The residue Res $\{\infty\}$ at infinity can be calculated as follows.

Let $\zeta^{-1} = \lambda$; then

$$\frac{1}{2\pi i} \int \lambda \frac{d}{d\lambda} \ln \vartheta \, d\lambda = -\frac{1}{2\pi i} \int \lambda \frac{d}{d\zeta} \ln \vartheta \, d\zeta . \tag{4.4.41}$$

Noting that the direction of ζ integration is the reverse of λ integration, we have

$$\text{Res}\{\infty\} = \text{Res}\left\{\lambda \frac{d}{d\lambda} \ln \vartheta \Big|_{\lambda=\infty}\right\} = \text{Res}\left\{\lambda \frac{d}{d\zeta} \ln \vartheta \Big|_{\zeta=0}\right\}. \tag{4.4.42}$$

However, since $d/d\zeta = -(1/\zeta^2)d/d\lambda$, if we use the notation

$$D_l = \frac{\partial}{\partial u_l} \tag{4.4.43}$$

for the upper sheet, we have

$$\frac{d}{d\zeta} \ln \vartheta(\boldsymbol{u} - \bar{\boldsymbol{c}}) = -\frac{1}{\zeta^2} \sum_{l=1}^{g} \frac{du_l}{d\lambda} D_l \ln \vartheta(\boldsymbol{u} - \bar{\boldsymbol{c}})$$

$$= -\frac{1}{\zeta^2} \sum_{l=1}^{g} \sum_{j=0}^{g-1} c_{lj} \frac{\lambda^j}{\sqrt{R(\lambda)}} D_l \ln \vartheta(\boldsymbol{u} - \bar{\boldsymbol{c}})$$

$$= -\frac{1}{\zeta^2} \sum_{l=1}^{g} \frac{c_{l,g-1}\lambda^{g-1} + c_{l,g-2}\lambda^{g-2} + \cdots}{\sqrt{\prod_{j=1}^{2g+2}(\lambda - \lambda_j)}} D_l \ln \vartheta(\boldsymbol{u} - \bar{\boldsymbol{c}})$$

$$= -\sum_{l=1}^{g} [c_{l,g-1} + O(\zeta)] D_l \ln \vartheta(\boldsymbol{u} - \bar{\boldsymbol{c}}) \tag{4.4.44}$$

for small ζ. Thus

$$\lim_{\lambda \to \infty} \lambda \frac{d}{d\zeta} \ln \vartheta(\boldsymbol{u} - \bar{\boldsymbol{c}}) = -\lim_{\zeta \to 0} \frac{1}{\zeta} \sum_{l=1}^{g} c_{l,g-1} D_l \ln \vartheta(\boldsymbol{u} - \bar{\boldsymbol{c}}) \tag{4.4.45}$$

and the residue at ∞ is given by (4.4.42) as

$$\mathrm{Res}\,\{\infty\} = -\sum_{l=1}^{g} c_{l,g-1} D_l \ln \vartheta[\boldsymbol{u}(\infty) - \bar{\boldsymbol{c}}]. \tag{4.4.46a}$$

For the lower sheet the sign of $\sqrt{R(\lambda)}$ is different, and we have

$$\mathrm{Res}\,\{\infty'\} = +\sum_{l=1}^{g} c_{l,g-1} D_l \ln \vartheta[\boldsymbol{u}(\infty') - \bar{\boldsymbol{c}}]. \tag{4.4.46b}$$

Therefore (4.4.39, 40) give

$$\sum_{j=1}^{g} \lambda(P_j) = \sum_{l=1}^{g} \int_{\alpha_l} \lambda\omega_l + \sum_{l=1}^{g} c_{l,g-1} D_l \ln \frac{\vartheta[\boldsymbol{u}(\infty) - \bar{\boldsymbol{c}}]}{\vartheta[\boldsymbol{u}(\infty') - \bar{\boldsymbol{c}}]}. \tag{4.4.47}$$

Problem 4.2. Show that $\tau_{jk} = \tau_{kj}$.
Hint: Use (4.4.7).

4.5 Solution to the Inverse Problem

Let

$$\sum_{j=1}^{g} u_l(P_j) = X_l; \tag{4.5.1}$$

then from (4.4.38a) we have

$$\bar{c}_l = X_l + K_l - \sum_{j=1}^{g} \tau_{lj} \tag{4.5.2}$$

where u_l is given by (4.4.21a). Therefore for a given X_l, the solution to the Jacobi inversion problem

$$\sum_{j=1}^{g} \int_{P_0}^{P_j} \omega_l = X_l \tag{4.5.3}$$

is given as the zeros $P_1, P_2, \ldots, P_g$ of $\vartheta[\boldsymbol{u}(P) - \bar{\boldsymbol{c}})]$ with $\bar{\boldsymbol{c}}$ specified by (4.5.2).

Equation (4.3.25) can be written as

$$\sum_{j=1}^{g} \int_{\mu_0}^{\mu_j(k)} \omega_l = X_l(k) \tag{4.5.4}$$

where

$$X_l(k) = k \int_{\infty'}^{\infty} \omega_l + \sum_{j=1}^{g} \int_{\mu_0}^{\mu_j(0)} \omega_l - \sum_{j=1}^{g} n_j \tau_{lj} + m_l . \tag{4.5.5}$$

Therefore, if the last term of (4.5.2) is absorbed in $X_l(k)$, or in the third term on the right-hand side of (4.5.5), $\mu_1(k), \mu_2(k), \ldots, \mu_g(k)$ are given as the values of λ at the zeros $P = P_j(j = 1,2, \ldots, g)$ of $\vartheta[\boldsymbol{u}(P) - X(k) - K]$, or

$$\mu_j = \lambda(P_j) . \tag{4.5.6}$$

Thus, by (4.4.47) we have

$$\sum_{j=1}^{g} \mu_j(k) = \sum_{l=1}^{g} \int_{\alpha_l} \lambda \omega_l + \sum_{l=1}^{g} c_{l,g-1} D_l \ln \frac{\vartheta[\boldsymbol{u}(\infty) - X - K]}{\vartheta[\boldsymbol{u}(\infty') - X - K]} \tag{4.5.7}$$

and by (4.5.5) we may write

$$u_l(\infty) - X_l - K_l = kc_l + d_l \tag{4.5.8}$$

with

$$\left. \begin{aligned} c_l &= \int_{\infty}^{\infty'} \omega_l \\ d_l &= -\sum_{j=1}^{g} \int_{\mu_0}^{\mu_j(0)} \omega_l + \sum_{j=1}^{g} n_j \tau_{lj} - m_l + \int_{\mu_0}^{\infty} \omega_l - K_l \end{aligned} \right\} . \tag{4.5.9}$$

By the periodicity (4.4.18) of ϑ, the second and third terms of d_l can be omitted. Further

$$u_l(\infty') = \int_{\mu_0}^{\infty'} \omega_l = \int_{\mu_0}^{\infty} \omega_l + \int_{\infty}^{\infty'} \omega_l = u_l(\infty) + c_l . \tag{4.5.10}$$

Therefore we have

$$\sum_{j=1}^{g} \mu_j(k) = \sum_{l=1}^{g} \int_{\alpha_l} \lambda \omega_l + \sum_{l=1}^{g} c_{l,g-1} D_l \ln \frac{\vartheta(k\boldsymbol{c} + \boldsymbol{d})}{\vartheta[(k + 1)\,\boldsymbol{c} + \boldsymbol{d}]} \qquad (4.5.11)$$

where the first term on the right-hand side is a constant. Inserting (4.5.11) into (4.2.32a) we finally obtain

$$b_{k+1} = \text{const.} - \sum_{l=1}^{g} c_{l,g-1} D_l \ln \frac{\vartheta(k\boldsymbol{c} + \boldsymbol{d})}{\vartheta[(k + 1)\,\boldsymbol{c} + \boldsymbol{d}]}. \qquad (4.5.12)$$

4.6 Time Evolution

The equations of motion for the lattice

$$\left.\begin{array}{l} \dot{Q}_n = P_n \\ \dot{P}_n = e^{-(Q_n - Q_{n-1})} - e^{-(Q_{n+1} - Q_n)} \end{array}\right\}, \qquad (4.6.1)$$

can be written as

$$\left.\begin{array}{l} \dot{a}_n = a_n(b_n - b_{n+1}) \\ \dot{b}_n = 2(a_{n-1}^2 - a_n^2) \end{array}\right\} \qquad (4.6.2)$$

with

$$a_n = \frac{1}{2}\, e^{-(Q_{n+1} - Q_n)/2}, \quad b_n = \frac{1}{2}\, P_n\,. \qquad (4.6.3)$$

We may also write them as

$$\dot{L} = BL - LB \qquad (4.6.4)$$

with the definition

$$\left.\begin{array}{l} (B\varphi)_n = -a_n \varphi(n + 1) + a_{n-1}\varphi(n - 1) \\ (L\varphi)_n = a_n \varphi(n + 1) + b_n \varphi(n) + a_{n-1}\varphi(n - 1) \end{array}\right\}. \qquad (4.6.5)$$

In addition, we impose the periodic conditions

$$Q_{n+N} = Q_n, \qquad P_{n+N} = P_n \qquad (n = -\infty, \cdots, \infty) \qquad (4.6.6)$$

which imply

$$A = \prod_{k=1}^{N} a_n = 2^{-N}. \qquad (4.6.7)$$

We consider $\varphi = \varphi(n)$ $(n = -\infty, ..., \infty)$ which satisfy (4.1.3), or the equation with a time-independent parameter λ:

$$L\varphi = \lambda\varphi \qquad (4.6.8a)$$

$$\dot{\lambda} = 0 . \qquad (4.6.8b)$$

Differentiating (4.6.8a) with respect to time we obtain

$$\dot{L}\varphi + L\dot{\varphi} = \lambda\dot{\varphi} . \qquad (4.6.9)$$

If we subtract (4.6.4) multiplied by φ from the above equation, we have

$$L(\dot{\varphi} - B\varphi) = \lambda(\dot{\varphi} - B\varphi) . \qquad (4.6.10)$$

Since this is of the same form as (4.6.8a), and if we let φ_1 and φ_2 be the fundamental solutions of (4.6.8a), we see that $\dot{\varphi}_1 - B\varphi_1$ and $\dot{\varphi}_2 - B\varphi_2$ are given as linear combinations of φ_1 and φ_2. For example

$$(\dot{\varphi}_1 - B\varphi_1)_n = \alpha\varphi_1(n) + \beta\varphi_2(n) \qquad (4.6.11a)$$

where α and β are certain constants (we could have such an argument in Sect. 3.2 for an infinite lattice. But as far as we assume that the lattice is at rest at infinity, by (3.2.8) we have $\varphi \to z^{\pm n}$ as $n \to \infty$ which implies $\alpha = \beta = 0$). We let $n = 0$ and 1 in (4.6.8a, 11a) and make use of the conditions $\varphi_1(0) = 1$, $\varphi_1(1) = 0$, $\varphi_2(0) = 0$, $\varphi_2(1) = 1$. Thus we get

$$\left.\begin{array}{l} (L\varphi_1)_0 = b_0 + a_{-1}\varphi_1(-1) = \lambda \\ (L\varphi_1)_1 = a_1\varphi_1(2) + a_0 = 0 \\ -(B\varphi_1)_0 = -a_{-1}\varphi_1(-1) = \alpha \\ -(B\varphi_1)_1 = a_1\varphi_1(2) - a_0 = \beta \end{array}\right\} \qquad (4.6.11b)$$

and therefore

$$\alpha = b_0 - \lambda, \qquad \beta = -2a_0 . \qquad (4.6.12)$$

Thus (4.6.11a) yields

$$\dot{\varphi}_1(n) = -a_n\varphi_1(n+1) + a_{n-1}\varphi_1(n-1) + (b_0 - \lambda)\varphi_1(n) - 2a_0\varphi_2(n) . \qquad (4.6.13)$$

In the same way, we have

$$(\dot{\varphi}_2 - B\varphi_2)_n = \bar{\alpha}\varphi_1(n) + \bar{\beta}\varphi_2(n) \qquad (4.6.14a)$$

so that

$$
\left.\begin{array}{l}
(L\varphi_2)_0 = a_0 + a_{-1}\varphi_2(-1) = 0 \\
(L\varphi_2)_1 = a_1\varphi_2(2) + b_0 = \lambda \\
-(B\varphi_2)_0 = a_0 - a_{-1}\varphi_2(-1) = \bar{\alpha} \\
-(B\varphi_2)_1 = a_1\varphi_2(2) = \bar{\beta}
\end{array}\right\} \tag{4.6.14b}
$$

and therefore

$$
\bar{\alpha} = 2a_0, \qquad \bar{\beta} = \lambda - b_0 . \tag{4.6.15}
$$

Thus (4.6.14a) yields

$$
\dot{\varphi}_2(n) = -a_n\varphi_2(n+1) + a_{n-1}\varphi_2(n-1) + 2a_0\varphi_1(n) + (\lambda - b_0)\varphi_2(n) . \tag{4.6.16}
$$

If we differentiate

$$
\Delta(\lambda) = \varphi_1(N) + \varphi_2(N+1) \tag{4.6.17}
$$

with respect to time, considering the periodic boundary conditions, we obtain

$$
\begin{aligned}
\dot{\Delta}(\lambda) &= \dot{\varphi}_1(N) + \dot{\varphi}_2(N+1) \\
&= -a_N\varphi_1(N+1) + a_{N-1}\varphi_1(N-1) + (b_N - \lambda)\varphi_1(N) - 2a_N\varphi_2(N) \\
&\quad -a_{N+1}\varphi_2(N+2) + a_N\varphi_2(N) + 2a_N\varphi_1(N+1) + (\lambda - b_N)\varphi_2(N+1) \\
&= (L\varphi_1 - \lambda\varphi_1)_N - (L\varphi_2 - \lambda\varphi_2)_{N+1} . \tag{4.6.18}
\end{aligned}
$$

Thus

$$
\dot{\Delta}(\lambda) = 0 . \tag{4.6.19}
$$

Namely, $\Delta(\lambda)$ does not depend on time.

Now, we put $n = N + 1$ in (4.6.5, 13) to obtain

$$
\left.\begin{array}{l}
a_1\varphi_1(N+2) + b_1\varphi_1(N+1) + a_0\varphi_1(N) = \lambda\varphi_1(N+1) \\
\dot{\varphi}_1(N+1) = -a_1\varphi_1(N+2) + a_0\varphi_1(N) + (b_0 - \lambda)\varphi_1(N+1) - 2a_0\varphi_2(N+1)
\end{array}\right\} \tag{4.6.20}
$$

Eliminating $\varphi_1(N+2)$, we get

$$
\dot{\varphi}_1(N+1) = 2a_0\varphi_1(N) + (b_0 + b_1 - 2\lambda)\varphi_1(N+1) - 2a_0\varphi_2(N+1) . \tag{4.6.21}
$$

The auxiliary spectrum μ satisfies $\varphi_1(N+1, \mu) = 0$. Therefore, by (4.1.41), we have

$$\dot{\varphi}_1(N + 1, \lambda)|_{\lambda=\mu} = 2a_0[\varphi_1(N, \mu) - \varphi_2(N + 1, \mu)]$$
$$= \pm 2a_0\sqrt{\Delta^2(\mu) - 4} \tag{4.6.22}$$

where we have used (4.1.30). On the right-hand side, we may rewrite

$$\sqrt{\Delta^2(\lambda) - 4} = A^{-1}Q(\lambda)\sqrt{R(\lambda)} \tag{4.6.23a}$$

with

$$Q(\lambda) = \prod_{l=g+1}^{N-1} [\lambda - \mu_l(0)], \quad R(\lambda) = \prod_{j=1}^{2g+2} (\lambda - \lambda_j). \tag{4.6.23b}$$

On the other hand, we may write

$$\varphi_1(N + 1, \lambda) = -a_0 A^{-1} \prod_{j=1}^{g} [\lambda - \mu_j(0)] \, Q(\lambda). \tag{4.6.24}$$

For $l \geq g + 1$, $\mu_l(0) = \lambda_{2l+1} = \lambda_{2l+2}$ is a constant. Therefore, if we differentiate (4.6.24) with respect to time, keeping λ constant, we have

$$\dot{\varphi}_1(N + 1, \lambda) = A^{-1} \, a_0 Q(\lambda) \sum_{j=1}^{g} \dot{\mu}_j(0) \prod_{l \neq j}' [\lambda - \mu_l(0)]$$
$$- \dot{a}_0 A^{-1} Q(\lambda) \prod_{j=1}^{g} [\lambda - \mu_j(0)]. \tag{4.6.25}$$

We then let $\lambda \to \mu_k(0)$ to obtain

$$\dot{\varphi}_1(N + 1, \lambda)|_{\lambda=\mu_k(0)} = A^{-1} \, a_0 Q[\mu_k(0)] \dot{\mu}_k(0) \prod_{l \neq k}' [\mu_k(0) - \mu_l(0)]. \tag{4.6.26}$$

Equating (4.6.22) with (4.6.26), and using (4.6.23a), we have

$$\dot{\mu}_k(0) = \mp 2\sqrt{R[\mu_k(0)]}/\prod_{l \neq k}' [\mu_k(0) - \mu_l(0)]. \tag{4.6.27}$$

We go back to (4.5.9) and ask for the time rate of change in d_l, which turns out to be

$$\dot{d}_l(t) = -\sum_{j=1}^{g} \dot{\mu}_j(0, t)\omega \, /d\lambda|_{\lambda=\mu_l(0)}$$
$$= -\sum_{j=1}^{g} \dot{\mu}_j(0, t) \frac{\displaystyle\sum_{s=0}^{g-1} c_{ls}\mu_j^s(0, t)}{\pm\sqrt{R[\mu_j(0, t)]}}. \tag{4.6.28}$$

On the right-hand side, using (4.6.27) we may write

$$\sum_{j=1}^{g} \frac{\dot{\mu}_j(0, t)\mu_j^s(0, t)}{\mp\sqrt{R[\mu_j(0, t)]}} = 2\sum_{j=1}^{g} \frac{\mu_j^s(0, t)}{\prod_{l \neq j}' [\mu_j(0, t) - \mu_j(0, t)]} \tag{4.6.29}$$

which can be simplified by the use of the formula (Lagrange's interpolation formula; cf. Appendix E)

$$\sum_{j=1}^{g}\frac{X_j^s}{\prod\limits_{l(\neq j)=1}^{g}(X_j - X_l)} = \begin{cases} 0 & (s < g - 1) \\ 1 & (s = g - 1) \end{cases}. \qquad (4.6.30)$$

Putting $X_j = \mu_j(0, t)$, from (4.6.29, 30) we have

$$\sum_{j=1}^{g}\frac{\dot{\mu}_j(0, t)\mu_j^s(0, t)}{\mp\sqrt{R[\mu_j(0, t)]}} = \begin{cases} 0 & (s < g - 1) \\ 2 & (s = g - 1) \end{cases}. \qquad (4.6.31)$$

Equations (4.6.27, 31) were first derived by *kac* and *van Moerbeke* [4.2]. Substituting (4.6.31) into (4.6.28) we have

$$\dot{d}_l(t) = - 2c_{l,g-1} \qquad (4.6.32)$$

or

$$d_l(t) = d_l(0) - 2c_{l,g-1}t . \qquad (4.6.33)$$

Thus, using (4.6.33) we can write (4.5.12) as

$$b_n(t) = \text{const.} - \sum_{l=1}^{g} c_{l,g-1}D_l \ln \frac{\vartheta[nc + d(t)]}{\vartheta[(n + 1) c + d(t)]} \qquad (4.6.34)$$

where we note from (4.6.33) that

$$2c_{l,g-1}D_l = -\frac{d}{dt} . \qquad (4.6.35)$$

Since $b_n = P_n/2 = \dot{Q}_n/2$ we have the final results

$$P_n(t) = \bar{P}_0 + \frac{d}{dt} \ln \frac{\vartheta(nc - c't + \delta')}{\vartheta[(n + 1) c - c't + \delta']} \qquad (4.6.36)$$

$$Q_n(t) = \bar{Q}_n(0) + \bar{P}_0 t + \ln \frac{\vartheta(nc - c't + \delta')}{\vartheta[(n + 1) c - c't + \delta']}, \qquad (4.6.37)$$

where $\bar{P}_0$ and $\bar{Q}_n$'s are some constants. These results were first obtained by *Date* and *Tanaka* [4.3]. In these formulas

$$c_l = \int_{\infty'}^{\infty}\omega_l, \quad c_l' = 2c_{l,g-1} \qquad (4.6.38)$$

and $\delta' = (\delta, ..., \delta_g')$ are phase constants determined by the initial conditions. c_l, c_l', δ_l are all purely imaginary as will be seen in a concrete example of a cnoidal wave in the next section. By definition, c_l and c'_l are determined by λ_1, λ_2, ..., λ_g and as the latter are determined by the initial conditions, so are all these quantities.

Further, we may write

$$Q_n = S_n - S_{n+1} \tag{4.6.39}$$

with

$$S_n = f + \ln \vartheta(nc - c't + \delta') \tag{4.6.40}$$

where f can be a certain function of n and t, but it must satisfy some conditions imposed on S_n. For example, we may have

$$f = \bar{f}(t) + n\bar{P}_0 t + Bn^2 \tag{4.6.41}$$

where $\bar{f}(t)$ depends only on t and B is a constant.

As already mentioned, coefficients $c_{l,k}$ of ω_l are imaginary, as are c_l and c'_l. Thus (4.6.40) is a ϑ function with imaginary arguments. In this case τ_{jk} is also pure imaginary. By Jacobi's imaginary transformation applied to the multidimensional ϑ function (cf. Appendix F)

$$\vartheta(u|\tau) = \frac{i^{g/2}}{|\tau|^{1/2}} e^{-\pi i u \cdot \tau^{-1} u} \vartheta(\tau^{-1} u| - \tau^{-1}), \tag{4.6.42}$$

where $|\tau| = \det \tau$ and τ^{-1} is the inverse matrix of τ, we can rewrite (4.6.40) in terms of a function with real arguments in such a way that

$$S_n = F + \ln \vartheta(\kappa n - \beta t + \delta| - \tau^{-1}). \tag{4.6.43}$$

F is some function similar to f in (4.6.41), and κ, β and δ are all real constants with g components such that

$$\kappa = \tau^{-1} c, \qquad \beta = \tau^{-1} c', \qquad \delta = \tau^{-1} \delta'. \tag{4.6.44}$$

4.7 A Simple Example (A Cnoidal Wave)

I) We apply the results obtained in the preceding sections to the simplest case where $g = 1$. As is shown in Fig. 4.9 we assume four roots such that

$$\lambda_1 < \lambda_2 < \lambda_3 < \lambda_4, \tag{4.7.1}$$

and related branch cuts $[\lambda_1, \lambda_2]$ and $[\lambda_3, \lambda_4]$. We consider an Abelian differential

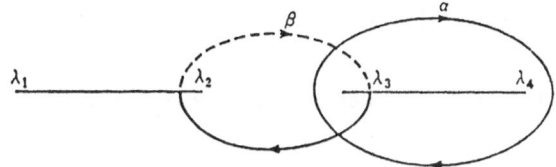

Fig. 4.9. The case of a cnoidal wave

$$\omega = \frac{c_0 d\lambda}{\sqrt{(\lambda - \lambda_1)(\lambda - \lambda_2)(\lambda - \lambda_3)(\lambda - \lambda_4)}} \tag{4.7.2}$$

with a constant c_0, and introduce an α-period (4.3.15) and a β-period (4.3.16) by

$$1 = \int_\alpha \omega = 2\int_{\lambda_3}^{\lambda_4} \frac{c_0 d\lambda}{\sqrt{(\lambda - \lambda_1)(\lambda - \lambda_2)(\lambda - \lambda_3)(\lambda - \lambda_4)}} \tag{4.7.3}$$

$$\tau_1 = \int_\beta \omega = 2\int_{\lambda_2}^{\lambda_3} \frac{c_0 d\lambda}{\sqrt{(\lambda - \lambda_1)(\lambda - \lambda_2)(\lambda - \lambda_3)(\lambda - \lambda_4)}}. \tag{4.7.4}$$

As customarily done for elliptic integrals, we make use of the transformation (linear transformation)

$$\lambda = \frac{\alpha s + \beta}{\gamma s + \delta}, \qquad s = \frac{\lambda\delta - \beta}{\alpha - \lambda\gamma} \tag{4.7.5}$$

where α, β, γ and δ are constants. We have

$$\int \omega = A_0 \int \frac{ds}{\sqrt{(1 - s/s_1)(1 - s/s_2)(1 - s/s_3)(1 - s/s_4)}}, \tag{4.7.6}$$

where

$$\left.\begin{array}{l} A_0 = \dfrac{(\alpha\delta - \beta\gamma)c_0}{\sqrt{(\beta - \lambda_1\delta)(\beta - \lambda_2\delta)(\beta - \lambda_3\delta)(\beta - \lambda_4\delta)}} \\[3mm] s_1 = \dfrac{\lambda_1\delta - \beta}{\alpha - \lambda_1\gamma}, \qquad s_2 = \dfrac{\lambda_2\delta - \beta}{\alpha - \lambda_2\gamma} \\[3mm] s_3 = \dfrac{\lambda_3\delta - \beta}{\alpha - \lambda_3\gamma}, \qquad s_4 = \dfrac{\lambda_4\delta - \beta}{\alpha - \lambda_4\gamma} \end{array}\right\}. \tag{4.7.7}$$

We introduce a constant κ such that $0 < \kappa < 1$, and let

$$s_1 = -\frac{1}{\kappa}, \qquad s_2 = -1, \qquad s_3 = 1, \qquad s_4 = \frac{1}{\kappa}. \tag{4.7.8}$$

We may solve these four equations for β, γ, δ, and κ, putting $\alpha = 1$. In particular, κ satisfies

$$\kappa^2 - 2\left[1 + \frac{2(\lambda_2 - \lambda_1)(\lambda_4 - \lambda_3)}{(\lambda_4 - \lambda_1)(\lambda_3 - \lambda_2)}\right]\kappa + 1 = 0 \tag{4.7.9}$$

which shows that we have one of the roots smaller than unity as we expected. We have thus

$$
\begin{aligned}
1 = \int_\alpha \omega = 2A_0 \int_1^{1/\kappa} \frac{ds}{\sqrt{(1 - s^2)(1 - \kappa^2 s^2)}} \\
= - 2iA_0 \int_1^{1/\kappa} \frac{ds}{\sqrt{(s^2 - 1)(1 - \kappa^2 s^2)}} \\
= - 2iA_0 K(\kappa'),
\end{aligned}
\tag{4.7.10}
$$

where $K(\kappa')$ is the complete elliptic integral of the first kind with the modulus $\kappa' = \sqrt{1 - \kappa^2}$. Further, we have

$$\tau_1 = \int_\beta \omega = 2A_0 \int_{-1}^1 \frac{ds}{\sqrt{(1 - s^2)(1 - \kappa^2 s^2)}} = 4A_0 K(\kappa). \tag{4.7.11}$$

Therefore

$$A_0 = i/2K(\kappa'), \qquad \tau_1 = i2K(\kappa)/K(\kappa'). \tag{4.7.12}$$

Since when κ varies from 0 to 1, $2K(\kappa)/K(\kappa')$ monotonously changes from 0 to ∞, we may introduce a new modulus k to write

$$\frac{2K(\kappa)}{K(\kappa')} = \frac{K(k)}{K(k')} \tag{4.7.13}$$

with $k' = \sqrt{1 - k^2}$, and write

$$\tau_1 = i\frac{K(k)}{K(k')}. \tag{4.7.14}$$

In this case the Riemann ϑ function reduces to the elliptic ϑ function ϑ_3, which can be written in terms of ϑ_0 since $\vartheta_3(v) = \vartheta_0(v + 1/2)$. We have

$$\vartheta_3(v \,|\, \tau_1) = \vartheta_3(v; q_1) = \sum_{m=-\infty}^\infty q_1^{m^2} e^{2\pi i m v}, \tag{4.7.15a}$$

where

$$q_1 = e^{i\pi\tau_1} = e^{-\pi K(k)/K(k')}. \tag{4.7.15b}$$

Using Jacobi's imaginary transformation for ϑ_3 [4.5]

$$\vartheta_3(v\,|\,\tau_1) = \sqrt{\frac{\mathrm{i}}{\tau_1}}\,\mathrm{e}^{-v^2\pi\mathrm{i}/\tau_1}\,\vartheta_3\left(\frac{v}{\tau_1}\,\bigg|\,-\frac{1}{\tau_1}\right), \tag{4.7.16}$$

we can write the solutions (4.6.34–40) in terms of a ϑ_3 function with the parameter

$$q = \mathrm{e}^{-\pi K(k')/K(k)} . \tag{4.7.17}$$

The result may be compiled by

$$S_n = f(t) + n\bar{P}_0 t + Bn^2 + \ln\vartheta_3\left(\frac{n}{N} - vt + \delta\,;q\right), \tag{4.7.18}$$

where N is the wavelength and v is the frequency. Equation (4.7.18) gives a lattice cnoidal wave (2.3.10) $[\vartheta_0(x) = \vartheta_3(x + 1/2)]$ with the waveform illustrated in Fig. 4.10.

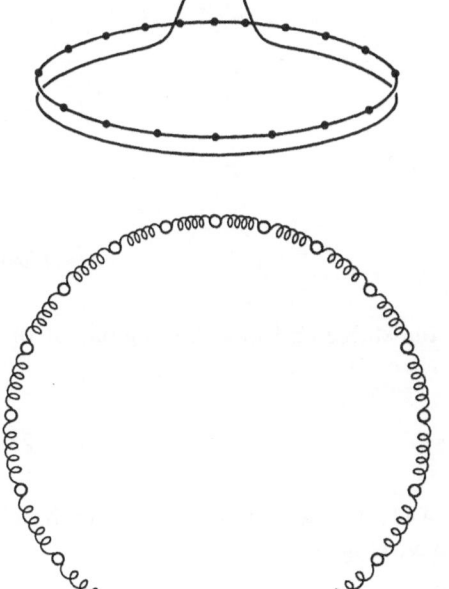

Fig. 4.10. A soliton of a periodic system (cnoidal wave)

II) We shall check some of the results of this chapter for the cnoidal wave with small amplitude. In this case we can use the continuum approximation, or the Korteweg-de Vries (KdV) equation. In this approximation the wave can be expressed by

$$u = 2r_n, \qquad r_n = Q_{n+1} - Q_n. \tag{4.7.19}$$

We express the wavelength of the cnoidal wave by the number of particles and let it be N. We assume the modulus k to be sufficiently small, so that the wave (initial form) is expressed by a sinusoidal wave

$$u \simeq -\varepsilon_0 \cos \frac{2\pi}{N} n, \qquad \varepsilon_0 = \frac{1}{4}\omega^2 k^2 = k^2 \sin^2 \frac{\pi}{N}. \tag{4.7.20}$$

By (3.11.23), the continuum version of $L\varphi = \lambda\varphi$ is the equation associated to the KdV equation, or

$$\frac{d^2\varphi}{dn^2} + (\lambda^{(\mathrm{KdV})} - u)\,\varphi = 0 \tag{4.7.21}$$

which reduces, for the sinusoidal wave, to the Mathieu equation [4.6]

$$\frac{d^2\varphi}{dz^2} + (\delta + \varepsilon \cos z)\,\varphi = 0, \tag{4.7.22}$$

where we have put

$$z = \frac{2\pi n}{N} \tag{4.7.23}$$

(the period of z is 2π), and

$$\delta = \left(\frac{N}{2\pi}\right)^2 \lambda^{(\mathrm{KdV})}, \qquad \varepsilon = \left(\frac{N}{2\pi}\right)^2 \varepsilon_0. \tag{4.7.24}$$

When ε is small, the boundaries of the stable and unstable regions of the Mathieu equation (4.7.22) are given by

$$\delta \simeq 0, \qquad \delta \simeq \frac{1}{4} \mp \frac{1}{2}\varepsilon \tag{4.7.25}$$

and there is a gap around $\delta \simeq 1/4$ (Fig. 4.11.a). If we rewrite these boundaries in terms of $\lambda^{(\mathrm{KdV})}$ as functions of ε_0, we have (Fig. 4.11b)

$$\lambda^{(\mathrm{KdV})} \simeq 0, \quad \lambda^{(\mathrm{KdV})} \simeq \left(\frac{\pi}{N}\right)^2 \mp \frac{1}{2}\varepsilon_0. \tag{4.7.26}$$

Further, the relation between $\lambda^{(\mathrm{KdV})}$ and the lattice spectrum $\lambda(L\varphi = \lambda\varphi)$ is

$$\left.\begin{array}{l} \lambda^{(\mathrm{KdV})} = k^2 \\[2mm] \lambda = \cos k = \cos \sqrt{\lambda^{(\mathrm{KdV})}} \end{array}\right\} \tag{4.7.27}$$

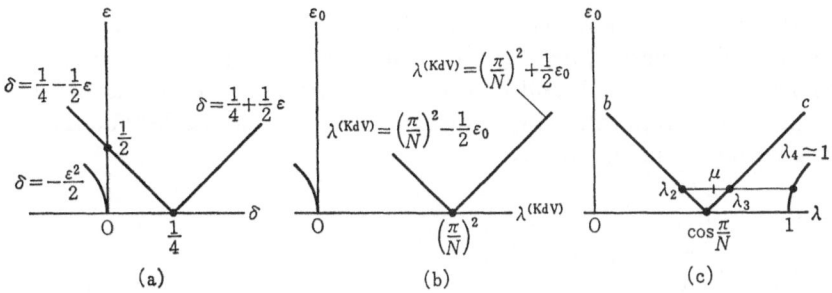

Fig. 4.11. Unstable regions and gaps of the Mathieu equation

for waves propagating to the right. Therefore, the boundaries of the lattice spectrum between stable and unstable regions are (Fig. 4.11c)

$$\lambda_d \simeq 1, \quad \binom{\lambda_b}{\lambda_c} \simeq \cos\sqrt{\left(\frac{\pi}{N}\right)^2 \mp \frac{\varepsilon_0}{2}} \simeq \cos\frac{\pi}{N} \pm \frac{\varepsilon_0}{4}\frac{\sin(\pi/N)}{\pi/N}. \tag{4.7.28}$$

The gap with width of about $\varepsilon_0/2$ and the narrow spectral band above it, from $\lambda \simeq \cos(\pi/N)$ to $\lambda \simeq 1$, are due to the cnoidal wave. The KdV equation takes up only the waves propagating in one direction, hence if we describe waves propagating to the right by the KdV approximation, the vicinity of $\lambda \simeq -1$ which corresponds to waves propagating to the left is disregarded. If we have lattice waves propagating to the left, we will have boundaries between stable and unstable regions near $\lambda \simeq -1$.

If $u = 0$ in (4.7.21), we have an eigenfunction

$$\varphi(n) = C \sin\frac{\pi}{N} n \tag{4.7.29}$$

belonging to the eigenvalue $\lambda^{(KdV)} = (\pi/N)^2$. Assuming, for simplicity, that $\pi/N \ll 1$ uses the continuum approximation, we normalize the eigenfunction in such a way that

$$\left.\begin{array}{c}\displaystyle\int_0^N |\varphi(n)|^2 dn = \frac{C^2 N}{2} = 1 \\[2mm] \therefore \quad C = \sqrt{\dfrac{2}{N}}\end{array}\right\}. \tag{4.7.30}$$

When we have the potential (4.7.20) shifted by k lattice points, the first-order perturbation theory yields the eigenvalue

$$\begin{aligned}\mu^{(KdV)}(k) &= \left(\frac{\pi}{N}\right)^2 - \frac{2}{N}\int_0^N \varepsilon_0 \cos\frac{2\pi}{N}(n+k)\sin^2\frac{\pi}{N} n \, dn \\ &= \left(\frac{\pi}{N}\right)^2 + \frac{\varepsilon_0}{2}\cos\frac{2\pi}{N} k. \end{aligned} \tag{4.7.31}$$

We note that $k = 0$ and $k = N/2$ give the boundaries between stable and unstable regions. If we vary k, $\mu^{(\text{KdV})}(k)$ oscillates between these boundaries.

We have defined λ_1, λ_2, λ_3 and λ_4 as the boundaries between stable and unstable regions for a given value of ε_0. If we denote by a, b, c, and d these bounndaries for sufficiently small ε_0, we have

$$
\left.
\begin{aligned}
\lambda_1 &\simeq a = -1 \\
\lambda_2 &\simeq b = \cos\frac{\pi}{N} - \frac{\varepsilon_0}{4}\frac{\sin(\pi/N)}{\pi/N} \\
\lambda_3 &\simeq c = \cos\frac{\pi}{N} + \frac{\varepsilon_0}{4}\frac{\sin(\pi/N)}{\pi/N} \\
\lambda_4 &\simeq d = 1
\end{aligned}
\right\}.
\tag{4.7.32}
$$

We may solve (4.7.7, 8) for α, β, γ, and δ, neglecting small quantities such as ε_0^2 and $\varepsilon_0\kappa$, to obtain

$$
\left.
\begin{aligned}
\alpha &= 4\kappa\cos\frac{\pi}{N}, \qquad \beta = 4\cos^2\frac{\pi}{N} \\
\gamma &= 4\kappa - \varepsilon_0\frac{\sin(\pi/N)}{\pi/N}, \qquad \delta = 4\cos\frac{\pi}{N}
\end{aligned}
\right\},
\tag{4.7.33}
$$

where κ is one of the roots of

$$
\kappa^2 - 4\left(\frac{\pi/N}{\varepsilon_0}\sin\frac{\pi}{N}\right)\kappa + 1 = 0
\tag{4.7.34}
$$

which is smaller than unity, or

$$
\kappa \simeq \frac{\varepsilon_0}{(4\pi/N)\sin(\pi/N)}.
\tag{4.7.35}
$$

The transformation (4.7.5) or the relation

$$
s = \frac{\delta\lambda - \beta}{\alpha - \gamma\lambda}
\tag{4.7.36}
$$

between s and λ is depicted in Fig. 4.12. $s = \pm\infty$ at

$$
\lambda_\infty = \frac{\alpha}{\gamma} = \frac{\kappa\cos(\pi/N)}{\kappa - \frac{\varepsilon_0}{4}\frac{\sin(\pi/N)}{\pi/N}} = \frac{1}{\cos(\pi/N)}
\tag{4.7.37}
$$

where we have used (4.7.35). $\lambda = \pm\infty$ at

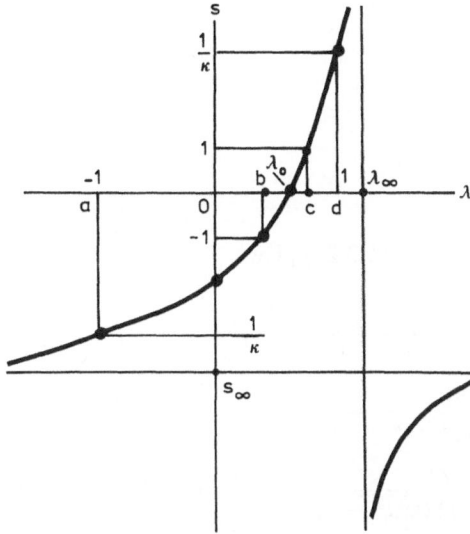

Fig. 4.12. The relation between s and λ

$$s_\infty = -\frac{\delta}{\gamma} = -\frac{\cos(\pi/N)}{\kappa - \dfrac{\varepsilon_0}{4}\dfrac{\sin(\pi/N)}{\pi/N}} = -\frac{\lambda_\infty}{\kappa} \tag{4.7.38}$$

and $s = 0$ at

$$\lambda_0 = \frac{\beta}{\delta} = \cos\frac{\pi}{N} + \frac{\varepsilon_0\kappa\dfrac{\sin(\pi/N)}{\pi/N}}{4\cos(\pi/N)}. \tag{4.7.39}$$

For $\lambda = \mu(k)$, where

$$\mu(k) = \cos\sqrt{\mu^{(\mathrm{KdV})}}(k) \simeq \cos\frac{\pi}{N} - \frac{\varepsilon_0}{4}\frac{\sin(\pi/N)}{\pi/N}\cos\frac{2\pi k}{N}, \tag{4.7.40}$$

we have

$$s[\mu(k)] \simeq \frac{-\cos\dfrac{\pi}{N}\cos\dfrac{2\pi k}{N}}{\cos\dfrac{\pi}{N} + \kappa\cos\dfrac{2\pi k}{N}} \simeq -\cos\frac{2\pi k}{N}. \tag{4.7.41}$$

Thus the left-hand side of (4.3.25) yields

$$\int_{\mu(0)}^{\mu(k)}\omega = A_0\int_{-1}^{-\cos(2\pi k/N)}\frac{-ds}{\sqrt{(1-s^2)(1-\kappa^2 s^2)}}$$

$$\simeq A_0 \int_1^{\cos(2\pi k/N)} \frac{ds}{\sqrt{1-s^2}}$$

$$= -A_0 \cos^{-1}\left(\cos\frac{2\pi k}{N}\right)$$

$$= -A_0 \frac{2\pi}{Nk} \quad (\mathrm{mod}\ A_0 2\pi). \tag{4.7.42}$$

On the other hand, the right-hand side of (4.3.25) gives

$$\int_{\lambda=d}^{\lambda=\infty} \omega = A_0 \left(\int_{1/\kappa}^\infty + \int_{-\infty}^{s\infty}\right) \frac{ds}{\sqrt{(s^2-1)\,(\kappa^2 s^2-1)}}$$

$$= A_0 \left(\int_{1/\kappa}^\infty + \int_{-s\infty}^\infty\right) \frac{ds}{\sqrt{(s^2-1)\,(\kappa^2 s^2-1)}}$$

$$= A_0 \left(2\int_{1/\kappa}^\infty - \int_{1/\kappa}^{-s\infty}\right) \frac{ds}{\sqrt{(s^2-1)\,(\kappa^2 s^2-1)}}$$

$$= A_0 \left[2K(\kappa) - \int_1^{\lambda\infty} \frac{(1/\kappa)dy}{\sqrt{(y^2/\kappa^2-1)\,(y^2-1)}}\right]$$

$$\simeq A_0 \left[2K(\kappa) - \int_1^{\lambda\infty} \frac{dy}{y\sqrt{y^2-1}}\right]$$

$$= A_0[2K(\kappa) - \arctan \sinh x_1] \tag{4.7.43}$$

with

$$x_1 = \cosh^{-1}\lambda_\infty. \tag{4.7.44}$$

Therefore, by (4.7.37), we have

$$\sinh x_1 = \sqrt{\lambda_\infty^2 - 1} = \tan\frac{\pi}{N} \tag{4.7.45}$$

and, using (4.5.9) and (4.7.43),

$$-c_1 = \int_{\infty'}^\infty \omega = 2\int_{\lambda=d}^\infty \omega \simeq -A_0\frac{2\pi}{N} + \tau_1. \tag{4.7.46}$$

Equations (4.7.42, 46) prove (4.3.25).

For small κ and k we can use

$$K(\kappa') = \frac{2}{\pi} K(\kappa) \ln\frac{4}{\kappa} \tag{4.7.47}$$

and write (4.7.13) as

$$2/\ln(4/\kappa) = 1/\ln(4/k) ,$$

which means

$$\left(\frac{4}{k}\right)^2 = \frac{4}{\kappa}$$

or

$$\kappa = \frac{k^2}{4} . \tag{4.7.48}$$

Since we are assuming that $\pi/N \ll 1$, comparing (4.7.20) with (4.7.35) we can verify (4.7.48). Thus, for small k, (4.7.17, 13, 47, 48) give

$$q \simeq \left(\frac{k}{4}\right)^2 \tag{4.7.49}$$

and we have the imaginary transformation for small q:

$$\vartheta_3 \simeq \sqrt{\frac{i}{\tau}}\, e^{-v^2\pi i/\tau_1}\left\{1 + 2q\, \cosh\left(2\pi \frac{K(k')}{K(k)} v\right)\right\}. \tag{4.7.50}$$

With respect to the argument v we may write

$$\left.\begin{array}{l} c_1 = A_0 \dfrac{2\pi}{N} = i\,\dfrac{\pi}{N}\dfrac{1}{K(\kappa')} \simeq i\,\dfrac{K(k)}{NK(k')} \\[2mm] v = nc_1 - \beta t + \delta' \end{array}\right\} \tag{4.7.51}$$

where we have used the fact that τ_1 in (4.7.46) can be omitted because it gives nothing to

$$\ln \vartheta_3 \simeq 2q\, \cos\left[2\pi\left(\frac{n}{N} - vt\right) + \delta\right]$$

$$+ \text{(terms linear with respect to } v^2\text{)} . \tag{4.7.52}$$

Here v and δ are some constants, cf. (4.6.42, 43). Thus we have

$$u \simeq 2\frac{d^2}{dn^2} S_n = - k^2\left(\frac{\pi}{N}\right)^2 \cos\left[2\pi\left(\frac{n}{N} - vt\right) + \delta\right] + \text{const.} \tag{4.7.53}$$

which coincides with (4.7.20), from which we have started.

If there is a soliton, we have an eigenvalue $\lambda^{(KdV)} < 1$ for a bound state. A cnoidal wave is a succession of solitons at equal intervals. When the modulus k

is close to unity the cnoidal wave consists of solitons with distinct peaks, and a very narrow band is below the continuous spectrum. Since this is similar in situation to a lattice soliton in an infinite system, the corresponding lattice spectrum is a band at $\lambda > 1$ (if the wave is propagating to the left, $\lambda < -1$). On the other hand, when the modulus k is very small, a cnoidal wave resembles a sinusoidal wave, and as is clear from (4.7.28) the spectral band is at $\lambda \simeq 1$. If the modulus of the cnoidal wave is increased to $k \simeq 1$, the unstable region of the Mathieu equation shifts to $\delta < 0$. Though the above approximation is not applicable when the modulus k approaches unity, it is clear that the unstable region for the lattice spectrum shifts to $|\lambda| > 1$. In Fig. 4.2 the curve represents the discriminant $\Delta(\lambda)$ as a function of λ, unstable regions ($\Delta^2 > 4$) at $|\lambda| \gtrsim 1$ are due to solitons, and unstable regions at $-1 < \lambda < 1$ are some ripplelike waves or radiations.

Such behavior of $\Delta(\lambda)$ was examined numerically by *Flaschka* et al., and, for periodic waves, they saw that the jth harmonic wave gave rise to an unstable region at the jth extremum (peak or valley) of the curve $\Delta(\lambda)$ vs λ. This fact can be understood as follows.

First, consider a lattice at rest. Then by (4.1.23, 24) $\Delta(\lambda) = 2\cos\alpha N$, $\lambda = \cos\alpha$, so that the roots of $\Delta(\lambda) = \pm 2$, or the positions of peaks and valleys are

$$\alpha_{(l)} = \frac{l\pi}{N} \qquad (l = 0,1,2, \cdots, N) \tag{4.7.54}$$

and the lth eigenvalue is

$$\lambda_{(l)} = \cos\frac{l\pi}{N} . \tag{4.7.55}$$

Now, we describe the effect of a sinusoidal wave with small amplitude by the KdV approximation assuming that the wave is propagating to the right. As we have seen in (3.11.6), the relation between the field u of the KdV equation and the displacement Q_n of the lattice is

$$u(n, t) = 2r_n(t) = 2(Q_{n+1} - Q_n) . \tag{4.7.56}$$

If we express the initial state of the jth harmonic wave by

$$u \simeq -\varepsilon_0 \cos\frac{j2\pi n}{N} , \tag{4.7.57}$$

then the eigenvalue equation

$$\frac{d^2\varphi}{dn^2} + (\lambda^{(\mathrm{KdV})} - u) \varphi = 0 \tag{4.7.58}$$

takes the form

$$\frac{d^2\varphi}{dz^2} + (\delta + \varepsilon \cos z)\,\varphi = 0\,, \tag{4.7.59}$$

with

$$z = \frac{j2\pi n}{N}, \qquad \delta = \left(\frac{N}{j2\pi}\right)^2 \lambda^{(\text{KdV})}, \qquad \varepsilon = \left(\frac{N}{j2\pi}\right)^2 \varepsilon_0\,. \tag{4.7.60}$$

The boundaries of stable and unstable regions of this Mathieu equation are

$$\delta \simeq 0, \qquad \delta \simeq \frac{1}{4} \mp \frac{1}{2}\varepsilon \tag{4.7.61}$$

and so the boundaries for the KdV equation are

$$\lambda^{(\text{KdV})} \simeq 0, \qquad \lambda^{(\text{KdV})} \simeq \left(\frac{j\pi}{N}\right)^2 \mp \frac{1}{2}\varepsilon_0\,. \tag{4.7.62}$$

Since the relation between the KdV spectrum $\lambda^{(\text{KdV})}$ and the lattice spectrum λ is (for waves propagating to the right)

$$\lambda = \cos\sqrt{\lambda^{(\text{KdV})}}\,, \tag{4.7.63}$$

the boundaries, including those of the gap, of the lattice spectrum are given as

$$\lambda_{(0)} \simeq 1$$
$$\lambda_{(j)}^{\pm} \simeq \cos\frac{j\pi}{N} \pm \frac{1}{4}\varepsilon_0\frac{\sin(j\pi/N)}{j\pi/N}$$
$$= \lambda_{(j)} \pm \frac{1}{4}\varepsilon_0\frac{\sin(j\pi/N)}{j\pi/N}\,. \tag{4.7.64}$$

This means that we have an unstable region at the jth extremum (peak or valley) of $\Delta(\lambda)$ (cf. Fig. 4.13).

The same can be said for the jth harmonic wave propagating to the left: when the amplitude is small the jth extremum (peak or valley) of $\Delta(\lambda)$ forms an unstable region $|\Delta(\lambda)| > 2$.

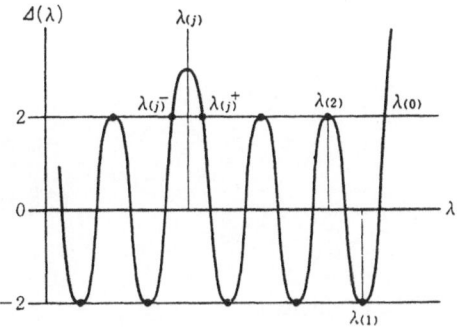

Fig. 4.13.   Unstable region due to the jth harmonic wave of small amplitude $[\Delta(\lambda) < 2]$

Problem 4.3. Verify (4.7.14) using the formulas

$$\int_b^x \frac{dx}{\sqrt{X}} = M \operatorname{sn}^{-1}\left[\sqrt{\frac{(c-a)(x-b)}{(c-b)(x-a)}}, \, k\right]$$

with the modulus k given by

$$k^2 = \frac{(c-b)(d-a)}{(d-b)(c-a)},$$

and

$$\int_c^x \frac{dx}{\sqrt{X}} = M \operatorname{sn}^{-1}\left[\sqrt{\frac{(d-b)(x-c)}{(d-c)(x-b)}}, \, k'\right]$$

with the modulus $k' = \sqrt{1-k^2}$, where

$$X = (x-d)(x-c)(x-b)(x-a)$$
$$M = 2/\sqrt{(d-b)(c-a)}$$
$$d > c > b > a.$$

4.8 Periodic System of Three-Particles

A three-particle system [4.8] is important because, though the simplest, it shows nearly all the characteristic features which many particle systems will exhibit. Therefore we describe it in some detail.

We first define the fundamental solutions φ_1 and φ_2 by

$$\left.\begin{array}{ll} \varphi_1(0) = 1, & \varphi_1(1) = 0 \\ \varphi_2(0) = 0, & \varphi_2(1) = 1 \end{array}\right\}. \tag{4.8.1}$$

We impose the cyclic boundary conditions

$$a_0 = a_3, \qquad b_0 = b_3. \tag{4.8.2}$$

Then we have the equations

$$\left.\begin{array}{l} a_0\varphi(0) + b_1\varphi(1) + a_1\varphi(2) = \lambda\varphi(1) \\ a_1\varphi(1) + b_2\varphi(2) + a_2\varphi(3) = \lambda\varphi(2) \\ a_2\varphi(2) + b_0\varphi(3) + a_0\varphi(4) = \lambda\varphi(3) \end{array}\right\} \tag{4.8.3}$$

for $\varphi = \varphi_1$ and φ_2, and the determinant of the coefficients

$$L^+ - \lambda I = \begin{pmatrix} b_1 - \lambda & a_1 & a_0 \\ a_1 & b_2 - \lambda & a_2 \\ a_0 & a_2 & b_0 - \lambda \end{pmatrix}. \tag{4.8.4}$$

As we know, the eigenvalues $\lambda_j(j = 1,2,3)$ of L^+ are independent of time. Solving (4.8.3) successively, we have

$$\varphi_1(2) = -\frac{a_0}{a_1}$$

$$\varphi_1(3) = -\frac{a_0}{a_1 a_2}(\lambda - b_2)$$

$$\varphi_1(4) = -\frac{\lambda - b_0}{a_0}\varphi_1(3) - \frac{a_2}{a_0}\varphi_1(2) = -\frac{(\lambda - b_2)(\lambda - b_0)}{a_1 a_2} + \frac{a_2}{a_1}$$

$$\varphi_2(2) = \frac{1}{a_1}(\lambda - b_1)$$

$$\varphi_2(3) = \frac{1}{a_1 a_2}(\lambda - b_1)(\lambda - b_2) - \frac{a_1}{a_2} \qquad\qquad (4.8.5)$$

$$\varphi_2(4) = \frac{1}{a_1 a_2 a_0}(\lambda - b_1)(\lambda - b_2)(\lambda - b_0)$$

$$\qquad\qquad - \frac{a_2}{a_0 a_1}(\lambda - b_1) - \frac{a_1}{a_2 a_0}(\lambda - b_0)$$

so that

$$\Delta(\lambda) = \varphi_1(3) + \varphi_2(4)$$

$$= \frac{1}{a_1 a_2 a_0}(\lambda - b_1)(\lambda - b_2)(\lambda - b_0)$$

$$\quad - \frac{a_2}{a_0 a_1}(\lambda - b_1) - \frac{a_0}{a_1 a_2}(\lambda - b_2) - \frac{a_1}{a_2 a_0}(\lambda - b_0), \tag{4.8.6}$$

and we have

$$\det(L^+ - \lambda I) = -a_1 a_2 a_0[\Delta(\lambda) - 2] \tag{4.8.7}$$

where $a_1 a_2 a_0 = 1/8$ and $\Delta(\lambda)$ is independent of time.

The auxiliary spectrum is given by

$$\varphi_1(4, \mu) = 0 \tag{4.8.8}$$

or

$$\frac{(\mu - b_2)(\mu - b_0)}{a_1 a_2} = \frac{a_2}{a_1}. \tag{4.8.9}$$

This is the same as the determinant equation

$$\det \begin{pmatrix} b_2 - \mu & a_2 \\ a_2 & b_0 - \mu \end{pmatrix} = 0 \tag{4.8.10}$$

which can be made using the right lower elements of $L^+ - \lambda I$, or

$$\mu^2 - (b_2 + b_0)\,\mu + (b_0 b_2 - a_2^2) = 0 \,. \tag{4.8.11}$$

The roots of this equation are

$$\left. \begin{aligned} \mu_1 &= \frac{b_2 + b_0 - \sqrt{(b_2 - b_0)^2 + 4a_2^2}}{2} \\ \mu_2 &= \frac{b_2 + b_0 + \sqrt{(b_2 - b_0)^2 + 4a_2^2}}{2} \end{aligned} \right\} \tag{4.8.12}$$

which oscillate with time (Fig. 4.14).

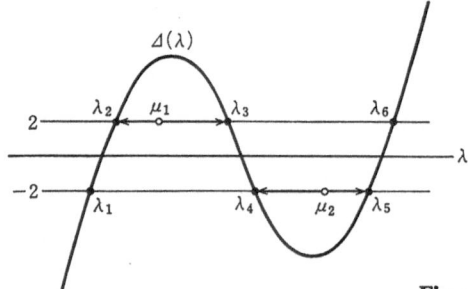

Fig. 4.14. $\Delta(\lambda)$ vs λ and auxiliary spectra μ_1 and μ_2 of a 3-particle system

In this case, for $\mu = \mu_1$ and μ_2, we have

$$\left. \begin{aligned} \varphi_1(3, \mu) &= -\frac{a_0}{a_1 a_2}\,(\mu - b_2) \\ \varphi_2(4, \mu) &= -\frac{a_1}{a_0 a_2}\,(\mu - b_0) \end{aligned} \right\}. \tag{4.8.13}$$

Therefore

$$\varphi_1(3, \mu)\,\varphi_2(4, \mu) = \frac{1}{a_2^2}\,(\mu - b_2)\,(\mu - b_0) = 1 \tag{4.8.14}$$

and we may write

$$\Delta(\mu) = \varphi_1(3, \mu) + \frac{1}{\varphi_1(3, \mu)}$$
$$= \pm 2 \cosh \ln |\varphi_1(3, \mu)| .$$ (4.8.15)

Further, since

$$[\varphi_1(3, \mu)]^2 - \Delta(\mu)\varphi_1(3, \mu) + 1 = 0 ,$$ (4.8.16)

we have

$$\varphi_1(3, \mu) = \frac{\Delta(\mu) \pm \sqrt{\Delta^2(\mu) - 4}}{2}$$ (4.8.17)

and

$$\Delta(\mu) = \varphi_1(3, \mu) + \varphi_2(4, \mu)$$
$$= -\frac{a_0}{a_1 a_2} (\mu - b_2) - \frac{a_1}{a_2 a_0} (\mu - b_0)$$
$$= -\frac{1}{a_1 a_2 a_0} [(a_0^2 + a_1^2) \mu - (a_0^2 b_2 + a_1^2 b_0)] .$$ (4.8.18)

Noting that

$$\left. \begin{array}{l} b_1 + b_2 + b_0 = \lambda_1 + \lambda_2 + \lambda_3 = \text{const.} \equiv b \\ \mu_1 + \mu_2 = b_2 + b_0 \\ \mu_2 - \mu_1 = \sqrt{(b_2 - b_0)^2 + 4a_2^2} \end{array} \right\},$$ (4.8.19)

we have

$$b_1^2 + b_2^2 + b_0^2 = (b - b_2 - b_0)^2 + b_2^2 + b_0^2$$
$$= 2(\mu_1^2 + \mu_2^2) + 2\mu_1\mu_2 - 2(\mu_1 + \mu_2) b - 2a_2^2 + b_2 .$$ (4.8.20)

Further, since we have

$$-\frac{[\Delta(\mu_1) - \Delta(\mu_2)]}{\mu_1 - \mu_2} = 8(a_1^2 + a_0^2) ,$$ (4.8.21)

the total energy of the system can be written as

$$E = 2(b_1^2 + b_2^2 + b_0^2) + 4(a_2^2 + a_1^2 + a_0^2)$$
$$= 4 \frac{-(1/8)[\Delta(\mu_1) - \Delta(\mu_2)] + \mu_1^3 - \mu_2^3}{\mu_1 - \mu_2} - 4(\mu_1 + \mu_2) b + 2b^2 .$$ (4.8.22)

The time rate of change of μ_1 is, by (4.8.12),

$$\dot{\mu}_1 = \frac{1}{2}\left[\dot{b}_2 + \dot{b}_0 - \frac{(b_2 - b_0)(\dot{b}_2 - \dot{b}_0) + 4a_2\dot{a}_2}{\sqrt{(b_2 - b_0)^2 + 4a_2^2}}\right]. \qquad (4.8.23)$$

We may use

$$\left.\begin{aligned}
\dot{b}_2 &= \frac{1}{2}\dot{p}_2 = 2(a_1^2 - a_2^2) \\
\dot{b}_0 &= \frac{1}{2}\dot{p}_0 = 2(a_2^2 - a_0^2) \\
\dot{a}_2 &= (b_2 - b_0)a_2
\end{aligned}\right\} \qquad (4.8.24)$$

to write

$$\dot{\mu}_1 = \frac{(a_1^2 - a_0^2)(\mu_2 - \mu_1) - (b_2 - b_0)(a_1^2 + a_0^2)}{\mu_2 - \mu_1}. \qquad (4.8.25)$$

However, using (4.8.14, 18, 19) we have

$$\sqrt{\Delta^2(\mu_1) - 4} = \frac{\pm 1}{a_1 a_2 a_0}[(a_0^2 - a_1^2)\mu_1 + (a_1^2 b_0 - a_0^2 b_2)]$$

$$= \frac{\pm 1}{2a_1 a_2 a_0}[(a_1^2 - a_0^2)(\mu_2 - \mu_1) - (b_2 - b_0)(a_1^2 + a_0^2)]. \quad (4.8.26)$$

Therefore, we obtain the result

$$\dot{\mu}_1 = \pm\frac{\sqrt{\Delta^2(\mu_1) - 4}}{4(\mu_1 - \mu_2)}. \qquad (4.8.27a)$$

The fundamental form of $\Delta(\mu)$ is determined by the initial condition, since it is independent of time [cf. (4.6.19)]. Thus, μ_1 oscillates between the roots of $\sqrt{\Delta^2(\lambda) - 4} = 0$, or between λ_1 and λ_2. Similarly, we have

$$\dot{\mu}_2 = \pm\frac{\sqrt{\Delta^2(\mu_2) - 4}}{4(\mu_2 - \mu_1)}, \qquad (4.8.27b)$$

μ_2 oscillates between λ_4 and λ_5. Therefore the system can be expressed by a set of points which oscillate between fixed boundaries.

Problem 4.4. Verify the following equations:

$$\Delta(\mu_1) + \sqrt{\Delta^2(\mu_1) - 4} = \frac{-2a_1^2}{a_1 a_2 a_0}(\mu_1 - b_0) = \Delta(\mu_2) - \sqrt{\Delta^2(\mu_2) - 4}$$

$$\Delta(\mu_1) - \sqrt{\Delta^2(\mu_1) - 4} = \frac{-2a_0^2}{a_1 a_2 a_0}(\mu_1 - b_2) = \Delta(\mu_2) + \sqrt{\Delta^2(\mu_2) - 4}.$$

5. Application of the Hamilton-Jacobi Theory

In this chapter, we begin with some simple examples to clarify the variable conjugate to the auxiliary spectrum. Then, by the help of such variables, we apply the Hamilton-Jacobi theory to the nonlinear lattice, and further derive action variables together with angle variables canonically conjugate to each other.

5.1 Canonically Conjugate Variables

Since the lattice with exponential interaction is integrable, we may apply the Hamilton-Jacobi theory and derive action and angle variables which are canonically conjugate. Action variables were first introduced for an infinite lattice [5.1], and then the application of the Hamiltonian-Jacobi theory was achieved for a periodic lattice. We shall be mainly concerned with periodic systems.

Let us begin with the simplest case of a 2-particle periodic lattice [5.2]. We refer to the center of mass coordinate so that the displacement of one of the particles is $q_1 = x$ and that of the other is $q_2 = -x$. Velocities are $\dot{q}_1 = \dot{x}$ and $\dot{q}_2 = -\dot{x}$, and the total energy is

$$E = \dot{x}^2 + e^{-2x} + e^{2x} \,. \tag{5.1.1}$$

Since the Lagrangean is $L = \dot{x}^2 - (e^{-2x} + e^{2x})$, the momentum conjugate to x is

$$p = \frac{\partial L}{\partial \dot{x}} = 2\dot{x} \tag{5.1.2}$$

and the Hamiltonian is

$$\mathscr{H}(x, p) = \frac{1}{4} p^2 + 2 \cosh 2x \,. \tag{5.1.3}$$

If we put

$$\mu = p/4, \quad \nu = -4x \,, \tag{5.1.4}$$

we may consider ν as the generalized momentum and μ as its conjugate coordinate (of course, we may take μ as the momentum and $-\nu$ as the coordinate). Then we have the Hamiltonian

$$\mathscr{H}(\mu, \nu) = 2 \cosh(\nu/2) + 4\mu^2 \tag{5.1.5}$$

and the equations of motion

$$\left.\begin{aligned} \dot{\mu} &= \frac{\partial \mathscr{H}}{\partial \nu} = \sinh \frac{\nu}{2} \\ \dot{\nu} &= -\frac{\partial \mathscr{H}}{\partial \mu} = -8\mu \end{aligned}\right\}. \tag{5.1.6}$$

Since we have

$$\ddot{\nu} = -8 \sinh \frac{\nu}{2}, \tag{5.1.7}$$

ν oscillates around $\nu = 0$, and μ also oscillates.

Next, consider a three-particle periodic system. As we have seen in Sect. 4.8, we have two auxiliary spectra μ_1 and μ_2 for this system, and they are subject to the equations of motion

$$\dot{\mu}_1 = \pm \frac{1}{4} \frac{\sqrt{\varDelta_1^2 - 4}}{\mu_1 - \mu_2}, \qquad \dot{\mu}_2 = \pm \frac{1}{4} \frac{\sqrt{\varDelta_2^2 - 4}}{\mu_2 - \mu_1} \tag{5.1.8a}$$

with

$$\varDelta_1 = \varDelta(\mu_1), \qquad \varDelta_2 = \varDelta(\mu_2). \tag{5.1.8b}$$

As we have seen in the preceding section, μ_1 and μ_2 oscillate respectively in the intervals (λ_2, λ_3) and (λ_4, λ_5). Compared with the case of the two-particle system, it is indicated that we may introduce momenta μ_1 and μ_2 canonically conjugate to ν_1 and ν_2, respectively, by assuming that $\sinh(\nu/2) \sim \sqrt{\varDelta^2(\mu) - 4}$.

To verify this assertion, we calculate the Poisson bracket

$$\begin{aligned} [\mu_1, \varDelta_1] = & \frac{\partial \mu_1}{\partial Q_1} \frac{\partial \varDelta_1}{\partial P_1} + \frac{\partial \mu_1}{\partial Q_1} \frac{\partial \varDelta_1}{\partial P_2} + \frac{\partial \mu_1}{\partial Q_0} \frac{\partial \varDelta_1}{\partial P_0} \\ & - \frac{\partial \mu_1}{\partial P_1} \frac{\partial \varDelta_1}{\partial Q_1} - \frac{\partial \mu_1}{\partial P_2} \frac{\partial \varDelta_1}{\partial Q_2} - \frac{\partial \mu_1}{\partial P_0} \frac{\partial \varDelta_1}{\partial Q_0}, \end{aligned} \tag{5.1.9}$$

where μ_1 and $\varDelta_1$ are functions of P_n, and Q_n through a_n and b_n ($n = 0,1,2$). By (4.8.12) and (4.8.18) we have

$$\left.\begin{aligned} \mu_1 &= \frac{b_2 + b_0 - \sqrt{(b_2 - b_0)^2 + 4a_2^2}}{2} \\ \mu_2 &= \frac{b_2 + b_0 + \sqrt{(b^2 - b_0)^2 + 4a_2^2}}{2} \end{aligned}\right\} \tag{5.1.10}$$

$$\left.\begin{array}{l} \varDelta_1 = -8[(a_0^2 + a_1^2)\,\mu_1 - (a_0^2 b_2 + a_1^2 b_0)] \\ \varDelta_2 = -8[(a_0^2 + a_1^2)\,\mu_2 - (a_0^2 b_2 + a_1^2 b_0)] \end{array}\right\}, \qquad (5.1.11)$$

and therefore

$$\mu_1 + \mu_2 = b_0 + b_2 \qquad\qquad\qquad\qquad (5.1.12a)$$

$$\mu_2 - \mu_1 = \sqrt{(b_2 - b_0)^2 + 4a_2^2} \qquad\qquad (5.1.12b)$$

$$\mu_1^2 + \mu_2^2 = b_2^2 + b_0^2 + 2a_2^2 \qquad\qquad\qquad (5.1.12c)$$

$$\mu_1\mu_2 = b_2 b_0 - a_2^2 \qquad\qquad\qquad\qquad (5.1.12d)$$

$$\varDelta_1 - \varDelta_2 = -8(a_0^2 + a_1^2)\,(\mu_1 - \mu_2)\,. \qquad (5.1.13)$$

After some calculation we obtain

$$\begin{aligned} \sqrt{\varDelta_1^2 - 4} &= 4[(a_1^2 - a_0^2)\,(\mu_2 - \mu_1) - (b_2 - b_0)(a_1^2 + a_0^2)] \\ &= 8[(a_0^2 - a_1^2)\mu_1 + (a_1^2 b_0 - a_0^2 b_2)] \end{aligned} \qquad (5.1.14)$$

and a similar formula for $\sqrt{\varDelta_2^2 - 4}$.

Then, noting that $a_2^2 = \{\exp[(Q_0 - Q_1)]\}/4$ and $b_2 = P_2/2$, we have

$$\left.\begin{array}{l} \dfrac{\partial \mu_1}{\partial P_1} = 0 \\[2.5ex] \dfrac{\partial \mu_1}{\partial Q_1} = 0 \\[2.5ex] \dfrac{\partial \mu_1}{\partial P_2} = \dfrac{\partial \mu_1}{\partial b_2}\dfrac{\partial b_2}{\partial P_2} \\[2.5ex] \qquad = \dfrac{1}{4}\left[1 - \dfrac{b_2 - b_0}{\sqrt{(b_2 - b_0)^2 + 4a_2^2}}\right] \\[3ex] \qquad = \dfrac{1}{4}\left(1 - \dfrac{b_2 - b_0}{\mu_2 - \mu_1}\right) \\[2.5ex] \dfrac{\partial \mu_1}{\partial Q_2} = \dfrac{\partial \mu_1}{\partial(a_2^2)}\dfrac{\partial(a_2^2)}{\partial Q_2} = \dfrac{-a_2^2}{\mu_2 - \mu_1} \\[2.5ex] \dfrac{\partial \mu_1}{\partial P_0} = \dfrac{\partial \mu_1}{\partial b_0}\dfrac{\partial b_0}{\partial P_0} = \dfrac{1}{4}\left(1 + \dfrac{b_2 - b_0}{\mu_2 - \mu_1}\right) \\[2.5ex] \dfrac{\partial \mu_1}{\partial Q_0} = \dfrac{\partial \mu_1}{\partial(a_2^2)}\dfrac{\partial(a_2^2)}{\partial Q_0} = \dfrac{a_2^2}{\mu_2 - \mu_1} \end{array}\right\}. \qquad (5.1.15)$$

Similarly as derivatives of $\varDelta_1$ we have

$$\left.\begin{aligned}
\frac{\partial \Delta_1}{\partial P_1} &= 0 \\[1ex]
\frac{\partial \Delta_1}{\partial Q_1} &= -8[-a_0^2(\mu_1 - b_2) + a_1^2(\mu_1 - b_0)] \\[1ex]
\frac{\partial \Delta_1}{\partial P_2} &= -2\left[(a_0^2 + a_1^2)\left(1 - \frac{b_2 - b_0}{\mu_2 - \mu_1}\right) - 2a_0^2\right] \\[1ex]
\frac{\partial \Delta_1}{\partial Q_2} &= 8\left[a_1^2(\mu_1 - b_0) + \frac{a_2^2(a_0^2 + a_1^2)}{\mu_2 - \mu_1}\right] \\[1ex]
\frac{\partial \Delta_1}{\partial P_0} &= -2\left[(a_0^2 + a_1^2)\left(1 + \frac{b_2 - b_0}{\mu_2 - \mu_1}\right) - 2a_1^2\right] \\[1ex]
\frac{\partial \Delta_1}{\partial Q_0} &= -8\left[a_0^2(\mu_1 - b_2) + \frac{a_0^2(a_0^2 + a_1^2)}{\mu_2 - \mu_1}\right]
\end{aligned}\right\} . \qquad (5.1.16)$$

Inserting these equations into (5.1.9) we obtain

$$[\mu_1, \Delta_1] = 2[(a_0^2 - a_1^2)\mu_1 + (a_1^2 b_0 - a_0^2 b_2)]$$
$$+ \frac{2}{\mu_2 - \mu_1}\{(b_2 - b_0)[a_1^2(\mu_1 - b_0) + a_0^2(\mu_1 - b_2)] + 2a_2^2(a_1^2 - a_0^2)\} . \qquad (5.1.17)$$

If we note that $\mu_1 = -(\mu_2 - \mu_1)/2 + (b_2 + b_0)/2$ by (5.1.12a), after using (5.1.12b), we get

$$[\mu_1, \Delta_1] = 2[(a_1^2 - a_0^2)\mu_1 + (a_1^2 b_0 - a_0^2 b_2)]$$
$$+ [-(b_2 - b_0)(a_1^2 + a_0^2) + (\mu_2 - \mu_1)(a_1^2 - a_0^2)] . \qquad (5.1.18)$$

Therefore (5.1.14) yields

$$[\mu_1, \Delta_1] = \frac{1}{4}\sqrt{\Delta_1^2 - 4} + \frac{1}{4}\sqrt{\Delta_1^2 - 4} = \frac{1}{2}\sqrt{\Delta_1^2 - 4} . \qquad (5.1.19)$$

Thus. noting the fact that $\Delta_1 \geq 2$ (cf. Fig. 4.13), if we introduce ν_1 by

$$\Delta_1 = 2\cosh\frac{\nu_1}{2} , \qquad (5.1.20)$$

(5.1.19) reduces to

$$[\mu_1, \Delta_1] = \sinh\frac{\nu_1}{2} . \qquad (5.1.21)$$

In general, if q and p are conjugate coordinate and momentum and if $f(p)$ is an arbitrary function of p, we have $[q, f(p)] = df(p)/dp$. Therefore (5.1.21) is equivalent to

$$[\mu_1, \nu_1] = 1 \qquad (5.1.22a)$$

and the momentum ν_1 conjugate to $\mu_2(\Delta_1 \geq 2)$ is given by

$$\nu_1 = 2 \ln \frac{\Delta_1 + \sqrt{\Delta_1^2 - 4}}{2}. \qquad (5.1.23a)$$

Similarly the momentum ν_2 conjugate to $\mu_2(\Delta_2 \leq -2)$ is given by

$$\nu_2 = 2 \ln \left| \frac{\Delta_2 - \sqrt{\Delta_2^2 - 4}}{2} \right|. \qquad (5.1.23b)$$

We can confirm the relations

$$\left.\begin{aligned}
[\mu_2, \nu_2] &= 1 \\
[\mu_1, \nu_2] &= [\mu_2, \nu_1] = 0 \\
[\mu_1, \mu_2] &= [\nu_1, \nu_2] = 0
\end{aligned}\right\} \qquad (5.1.22b)$$

which must hold for these conjugate variables. Here ν_1 is a function of μ_1 and does not depend on μ_2, and ν_2 is a function of μ_2 and does not depend on μ_1.

So far we considered two-particle and three-particle systems. We can extend the results to a many-particle periodic lattice [5.2, 3]. Let $\lambda_j^+(j = 1, 2, ..., N)$ be the eigenvalues of L^+ of (4.1.28) for a periodic lattice of N particles, or the roots of

$$\det(L^+ - \lambda I) = 0. \qquad (5.1.24)$$

Then the total energy of the lattice turns out to be

$$\begin{aligned}
E &= 2(b_1^2 + b_2^2 + \cdots + b_N^2) + 4(a_1^2 + a_2^2 + \cdots + a_N^2) \\
&= 2 \operatorname{tr}(L^+)^2 = 2 \sum_{j=1}^{N} (\lambda_j^+)^2.
\end{aligned} \qquad (5.1.25)$$

Since λ_j is the root of $\Delta(\lambda) = 2$, we may write

$$\Delta(\lambda) = A^{-1} \prod_{j=1}^{N} (\lambda - \lambda_j^+) + 2, \qquad (5.1.26)$$

which can be rewritten in a form of a power series of λ as

$$\Delta(\lambda) = \sum_{j=0}^{N} \lambda^j \beta_j. \qquad (5.1.27)$$

The coefficients β_j are independent of time, because $\Delta(\lambda)$ is time independent. In particular, we see from (5.1.26, 27) that

$$
\left.
\begin{aligned}
A\beta_N &= 1 \\
A\beta_{N-1} &= -\sum_{j=1}^{N} (\lambda^+)^j \\
&= -\operatorname{tr}\{L^+\} \\
&= -(b_1 + b_2 + \cdots + b_N) = -\frac{1}{2}P
\end{aligned}
\right\}, \tag{5.1.28}
$$

where P is the total momentum. Further we have

$$
\begin{aligned}
A\beta_{N-2} &= \sum_{i<j} \lambda_i^+ \lambda_j^+ = \frac{1}{2}\left(\sum_{j=1}^{N} \lambda_j^+\right)^2 - \frac{1}{2}\sum_{j=1}^{N} (\lambda_j^+)^2 \\
&= \frac{1}{8}P^2 - \frac{1}{4}E
\end{aligned} \tag{5.1.29}
$$

which is thus related to the total energy E. For simplicity, we choose a coordinate system for which the total momentum vanishes,

$$
P = 0, \qquad \beta_{N-1} = 0 \tag{5.1.30}
$$

Then we have

$$
E = -4A\beta_{N-2} \tag{5.1.31}
$$

We make use of Lagrange's interpolation formula

$$
\sum_{i=1}^{N-1} \frac{\mu_i^l}{P'(\mu_i)} = \begin{cases} 0 & (l < N-2) \\ 1 & (l = N-2) \end{cases} \tag{5.1.32a}
$$

with

$$
P(\mu) = \sum_{j=1}^{N-1} (\mu - \mu_j), \tag{5.1.32b}
$$

to have [5.2]

$$
\beta_{N-2} = \sum_{i=1}^{N-1} \frac{\beta_{N-2}\mu_i^{N-2} + \cdots + \beta_0}{P'(\mu_i)} = \sum_{i=1}^{N-1} \frac{\Delta(\mu_i) - \mu_i^N \beta_N}{P'(\mu_i)}. \tag{5.1.33}
$$

Therefore, if we rewrite the right-hand side of the last equation by using (5.1.27) for $\lambda = \mu_i$ [cf. the first equation of (5.1.28)] we obtain

$$
E = 4\sum_{i=1}^{N-1} \frac{-A\Delta(\mu_i) + \mu_i^N}{P'(\mu_i)} \tag{5.1.34}
$$

where we have assumed that there is no degeneracy, or that $\lambda_{2j} \neq \lambda_{2j+1}$, so that we have μ_i's from $i = 1$ to $i = N - 1$. If we have a double root ($\lambda_{2j} = \mu_j = \lambda_{2j+1}$) proper consideration is required. In principle, we may assume that all the degeneracies are removed by some small perturbation.

If N is odd (even) and i is odd (even, then $\Delta(\mu_i)$ is greater than 2, otherwise it is smaller than -2. Taking this fact into consideration we write

$$\Delta(\mu_i) = (-1)^{N-i} 2 \cosh \frac{\nu_i}{2} \tag{5.1.35}$$

and, using (5.1.34), define a function of (μ_i, ν_i) by

$$\mathscr{H}(\mu, \nu) = 4 \sum_{i=1}^{N-1} \frac{-A(-1)^{N-i} 2 \cosh(\nu_i/2) + \mu_i^N}{P'(\mu_i)}. \tag{5.1.36}$$

If we consider ν_i as the momentum conjugate to μ_i and set up the canonical equation of motion

$$\mu_i = \frac{\partial \mathscr{H}}{\partial \nu_i}, \tag{5.1.37}$$

it is shown that this actually coincides with the equation of motion for μ_i, and thus $\mathscr{H}$ is the Hamiltonian of the system. In fact, by (5.1.36), we have from (5.1.37)

$$\dot{\mu}_i = -\frac{4A(-1)^{N-i} \sinh(\nu_i/2)}{P'(\mu_i)}. \tag{5.1.38}$$

On the other hand, (5.1.35) yields

$$(-1)^{N-i} 2 \sinh \frac{\nu_i}{2} = \pm \sqrt{\Delta^2(\mu_i) - 4}. \tag{5.1.39}$$

Thus, we see that (5.1.38) definitely gives the equation of motion (4.6.27) for μ_i, or

$$\dot{\mu}_i = \pm \frac{2A\sqrt{\Delta^2(\mu_i) - 4}}{\prod'_{k \neq i}(\mu_i - \mu_k)}. \tag{5.1.40}$$

Further, by (5.1.35), the value of $\mathscr{H}$ (5.1.36) coincides with the energy E (5.1.34). Thus, $\mathscr{H}(\mu, \nu)$ of (5.1.36) is the Hamiltonian of the system, and ν_i and μ_i constitute conjugate variables. By (5.1.39, 35), we may write

$$\nu_i = 2 \ln \left| \frac{\Delta(\mu_i) \pm \sqrt{\Delta^2(\mu_i) - 4}}{2} \right| \tag{5.1.41}$$

where the double sign is $+$ for odd (even) i, and $-$ for even (odd) i, when N is odd (even). ν_i is a function of μ_i, and does not depend on other $\mu_j(j \neq i)$. When the initial conditions are given, the function $\Delta(\lambda)$ is settled and so we have the functions $\nu_i(\mu_i)$.

Problem 5.1. Similarly to the text, show that

$$[\mu_1, \Delta_2] = [\mu_2, \Delta_1] = 0$$
$$[\mu_2, \Delta_2] = -\sinh(\nu_2/2),$$

where $\Delta_2 = -2\cosh(\nu_2/2)$.

5.2 Action Variables

If we refer to the coordinate system for which $P = 0$, the Hamiltonian of the total system is, by (5.1.29), given as

$$\mathcal{H}(\mu_1, \cdots, \mu_{N-1}, \nu_1, \cdots, \nu_{N-1}) = E \tag{5.2.1}$$

where $E = -4A\beta_{N-2}$. According to the Hamilton-Jacobi theory, we transform the variables from (μ_i, ν_i) to (α_i, β_i) where β_i's have been defined in (5.1.27). We introduce as usual the transformation function

$$W(\mu_1, \cdots, \mu_{N-1}, \beta_0, \cdots, \beta_{N-3}, E, t)$$
$$= S(\mu_1, \cdots, \mu_{N-1}, \beta_0, \cdots, \beta_{N-3}, E) - Et, \tag{5.2.2}$$

which is to be determined in such a way that

$$\frac{\partial S}{\partial \mu_i} = \nu_i \qquad (i = 0, \cdots, N-1) \tag{5.2.3a}$$

$$\frac{\partial S}{\partial \beta_k} = \alpha_k \qquad (k = 0, \cdots, N-3) \tag{5.2.3b}$$

$$\frac{\partial S}{\partial E} = t + \alpha_{N-2}, \tag{5.2.3c}$$

where we have put $\beta_{N-1} = 0$ because of $P = 0$. In general, a canonical transformation with the transformation function W transforms the Hamiltonian to $K = \mathcal{H} + \partial W/\partial t$. The Hamilton-Jacobi transformation (5.2.3a–3c) is the one for which

$$K = \mathcal{H} + \frac{\partial W}{\partial t} = 0. \tag{5.2.4}$$

Thus, we have $\dot{\alpha}_k = \partial K/\partial \beta_k = 0$. Therefore, α_k which is conjugate to β_k is also time independent. Inserting (5.2.3a) into (5.2.1) we have

$$\mathscr{H}\left(\mu_1, \cdots, \mu_{N-1}, \frac{\partial S}{\partial \mu_1}, \cdots, \frac{\partial S}{\partial \mu_{N-1}}\right) = E \qquad (5.2.5)$$

which is the Hamilton-Jacobi equation for S.

Comparing (5.2.3a) with (5.1.41) we see that

$$\frac{\partial S}{\partial \mu_i} = v_i = 2 \ln\left|\frac{\Delta(\mu_i) \pm \sqrt{\Delta^2(\mu_i) - 4}}{2}\right|. \qquad (5.2.6)$$

Therefore, S is separated as

$$S = \sum_{i=1}^{N-1} S_i \qquad (5.2.7a)$$

with

$$S_i = \int v d\mu = 2 \int_{\lambda_{2i}}^{\mu_i} \ln\left|\frac{\Delta(\mu) \pm \sqrt{\Delta^2 - 4}}{2}\right| d\mu , \qquad (5.2.7b)$$

where $\Delta(\mu)$ depends on $\beta_0, \ldots, \beta_{N-2}$ by (5.1.27). Further, by (5.2.3a) and (5.1.27) we have

$$\begin{aligned}
\alpha_k &= \frac{\partial S}{\partial \beta_k} \\
&= 2 \sum_{i=1}^{N-1} \int_{\lambda_{2i}}^{\mu_i} \frac{1}{\sqrt{\Delta^2(\mu) - 4}} \frac{\partial \Delta(\mu)}{\partial \beta_k} d\mu \\
&= 2A \sum_{i=1}^{N-1} \int_{\lambda_{2i}}^{\mu_i} \frac{\mu^k}{\sqrt{R(\mu)}} d\mu
\end{aligned} \qquad (5.2.8)$$

and therefore, cf. (4.6.31)

$$\dot{\alpha}_k = 2A \sum_{i=1}^{N-1} \frac{\mu_i^k \dot{\mu}_i}{\sqrt{R(\mu_i)}} = 0 \qquad (k = 0,1, \cdots, N-3) \qquad (5.2.9a)$$

which confirms that $\alpha_k(k = 0,1, \ldots, N-3)$ is constant. The constants α_k can be chosen to be identically zero by adding β_k appropriately to S_i. In particular for $k = N - 2$, a similar calculation gives

$$\frac{d}{dt} \frac{\partial S}{\partial E} = 0 . \qquad (5.2.9b)$$

Therefore, by (5.2.3c) we see that $\dot{\alpha}_{N-2} = 0$, and α_{N-2} is a constant which can be chosen to be null.

The momentum ν_i is a function of μ_i and independent of other $\mu_j (j \neq i)$. In other words, ν_i's are mutually separated. By integrating it over a period, we have the action variable [5.1, 2]

$$
\begin{aligned}
J_i(\beta_0, \cdots, \beta_{N-1}) &= \oint \nu_i d\mu_i \\
&= 4 \int_{\lambda_{2i}}^{\lambda_{2i+1}} \ln \left| \frac{\Delta(\mu) \pm \sqrt{\Delta^2(\mu) - 4}}{2} \right| d\mu \\
&\quad (i = 1, 2, \cdots, N - 2),
\end{aligned}
\tag{5.2.10}
$$

where λ_{2i} and λ_{2i+1} are functions of $\beta_0, ..., \beta_{N-2}$ by (5.1.22a). Integration in (5.2.10) is limited between λ_{2j} and λ_{2j+1} where $|\Delta(\mu)| \geq 2$. The motion in the lattice is thus completely separated.

The angle variable w_i conjugate to the action variable J_i is given by

$$
w_i = \frac{\partial S}{\partial J_i}.
\tag{5.2.11}
$$

If we choose $\alpha_k = 0$ ($k = 1, ..., N - 2$), the angle variable turns out to be

$$
\begin{aligned}
w_i &= \sum_{k=0}^{N-2} \frac{\partial S}{\partial \beta_k} \frac{\partial \beta_k}{\partial J_i} \\
&= \frac{\partial E}{\partial J_i} t
\end{aligned}
\tag{5.2.12}
$$

by (5.2.4), where we have used, for $k = N - 2$, the fact that $E = -4A\beta_{N-2}$. Thus, $\partial E / \partial J_i$ is the frequency of oscillation, and if we denote the period by T_i, we have

$$
w_i = \frac{t}{T_i}.
\tag{5.2.13}
$$

As an example, let us consider a three-particle periodic lattice with one of the particles $n = 2$ at rest. Let, initially, $P_2 = 0$, $P_3 = -P_1$, and $Q_1 = Q_2 = Q_3 = 0$, so that $b_3 = -b_1$ and $a_1 = a_2 = a_3 = 1/2$ at $t = 0$. Therefore, we have

$$
\begin{aligned}
\Delta(\lambda) &= 8\lambda(\lambda^2 - b_1^2) - 6\lambda \\
&= 8\lambda^3 - (6 + 8b_1^2)\lambda.
\end{aligned}
\tag{5.2.14}
$$

If all the particles are at rest, then $b_1 = 0$ and the roots of $\Delta(\lambda) = \pm 2$ are ± 1 and $\pm 1/2$. The curve $\Delta(\lambda)$ vs λ has a point symmetry about $\lambda = 0$ when $b_2 = 0$ even if $b_1 \neq 0$.

The total energy is known from the initial data to be

$$E = \frac{1}{2}(P_1^2 + P_3^2) = 4b_1^2 . \tag{5.2.15}$$

Thus we may write

$$\Delta(\lambda) = 2[4\lambda^3 - (3 + E)\lambda] . \tag{5.2.16}$$

Denoting by λ_2 and λ_3 the roots of $\Delta(\lambda) = 2$, we have the action variable for μ_1 as

$$J_1 = 4 \int_{\lambda_2}^{\lambda_3} d\lambda \, \ln \frac{\Delta(\lambda) + \sqrt{\Delta^2 - 4}}{2} . \tag{5.2.17}$$

Therefore

$$\delta J_1 = 4\delta E \int_{\lambda_2}^{\lambda_3} d\lambda \frac{1 + \Delta/\sqrt{\Delta^2 - 4}}{\Delta + \sqrt{\Delta^2 - 4}} \frac{d\Delta}{dE}$$

$$= 4\delta E \int_{\lambda_2}^{\lambda_3} \frac{-2\lambda d\lambda}{\sqrt{\Delta^2 - 4}} . \tag{5.2.18}$$

In this case, we have the same change in J_2, so that $\delta J_1 = \delta J_2$, and therefore $\delta E = (\partial E/\partial J_1)\,\delta J_1 + (\partial E/\partial J_2)\,\delta J_2 = 2(\partial E/\partial J_1)\,\delta J_1$. Thus, the period T is given by

$$\frac{1}{T} = \frac{\partial E}{\partial J_1} = \left(16 \int_{\lambda_2}^{\lambda_3} \frac{-\lambda d\lambda}{\sqrt{\Delta^2 - 4}}\right)^{-1} . \tag{5.2.19}$$

The above is a very simple motion, and can be described by the equation of motion

$$\frac{d^2 Q}{dt^2} = e^{-Q} - e^{2Q} . \tag{5.2.20}$$

Integrating once, we have

$$\left(\frac{dQ}{dt}\right)^2 = E + 3 - (2e^{-Q} + e^{2Q}) . \tag{5.2.21}$$

Therefore the period is

$$T = 2 \int_{Q_-}^{Q_+} \frac{dQ}{\sqrt{E + 3 - 2e^{-Q} - e^{2Q}}}$$

$$= 2 \int_{Q_-}^{Q_+} \frac{e^{-Q} dQ}{\sqrt{-2e^{-3Q} + (E + 3)\,e^{-2Q} - 1}}$$

$$= 2 \int_{x_-}^{x_+} \frac{dx}{\sqrt{2x^3 + (E + 3)\,x^2 - 1}} , \tag{5.2.22}$$

where Q_- and Q_+, or x_- and x_+, are the zeros of the integrand, which are the turning points of the oscillation.

We can easily prove that (5.2.19, 22) have the same meaning. In fact, from (5.2.19) we have

$$
\begin{aligned}
T &= 16 \int_{\lambda_3}^{\lambda_2} \frac{\lambda d\lambda}{\sqrt{4\lambda^2[4\lambda^2 - (E + 3)]^2 - 4}} \\
&= 4 \int_{\lambda_3^2}^{\lambda_2^2} \frac{d(\lambda^2)}{\sqrt{4\lambda^2[2\lambda^2 - (E + 3)/2]^2 - 1}}.
\end{aligned}
\tag{5.2.23}
$$

If we put

$$
y = 2\lambda^2 - \frac{E + 3}{2},
\tag{5.2.24}
$$

we have

$$
T = 2 \int_{y_-}^{y_+} \frac{dy}{\sqrt{2y^3 + (E + 3)\, y^2 - 1}}
\tag{5.2.25}
$$

which is the same as (5.2.22).

6. Recent Advances in the Theory of Nonlinear Lattices

In this chapter we present some of the recent studies in the theory of nonlinear waves which bear on the problems of nonlinear lattices. The topics include, among others, a discussion of the problems of integrability, the generalized lattices and the Bethe ansatz. Some numerical results are also presented. Owing to the elaborate nature of these studies, it has not always been possible to give an adequately detailed account of them, but attempts have been made to give an indication of the wider aspects and of the methods employed. Since the description is frequently concerned with the so-called Toda lattice, this is frequently referred to by the abbreviation TL.

6.1 The KdV Equation as a Limit of the TL Equation

Both the KdV and the TL equations are integrable, and the form of the solutions as well as the method of solution of these equations are quite similar, indicating some close relation between them. In fact, we can show that the TL equation can be transferred to the KdV equation through the integrable regime [6.1]. The KdV is a limit of the TL equation.

To show this, we write the TL equation as

$$\frac{d^2}{dt^2} \log (1 + V_n) = V_{n+1} + V_{n-1} - 2 V_n , \tag{6.1.1}$$

and introduce a positive parameter h $(0 < h \leqslant 1)$ to write

$$t = \frac{\tau}{h^3} , \qquad V_n = h^2 u_n(\tau) . \tag{6.1.2}$$

Then (6.1.1) is written as

$$\frac{d^2}{d\tau^2} \log (1 + h^2 u_n) = \frac{1}{h^4} (u_{n+1} + u_{n-1} - 2 u_n) . \tag{6.1.3}$$

Further, we introduce a coordinate

$$x = h n - \left(\frac{1}{h^2} - h^2 \right) \tau . \tag{6.1.4}$$

For $0 < h < 1$, the coordinate x moves to the right relative to the lattice. We change the variables from (n, t) to (x, τ), and write

$$u(x, \tau) = u_n(\tau) \ . \tag{6.1.5}$$

The $d^2/d\tau^2$ on the l.h.s. of (6.1.3) means the derivative keeping n constant. Because of (6.1.4), we have in general

$$\frac{d}{d\tau} f(u_n(\tau)) = \left\{ \frac{\partial}{\partial \tau} - \left(\frac{1}{h^2} - h^2 \right) \frac{\partial}{\partial x} \right\} f(u(x, \tau)) \ . \tag{6.1.6}$$

Therefore (6.1.3) is written as

$$\left\{ \frac{\partial}{\partial \tau} - \left(\frac{1}{n^2} - h^2 \right) \frac{\partial}{\partial x} \right\}^2 \log \{ 1 + h^2 u(x, \tau) \}$$

$$= \frac{1}{h^4} \{ u(x + h, \tau) + u(x - h, \tau) - 2u(x, \tau) \} \ . \tag{6.1.7}$$

If we put $h = 1$, (6.1.7) reduces to the usual TL. But for $h \neq 1$, (6.1.7) is still the TL, and remains in the integrable regime, for we have only applied rescaling of variables and a coordinate transformation without any approximation.

We now take the limit of $h \to 0_+$. Then we see that (6.1.7) goes over to

$$\frac{\partial}{\partial x} \left\{ -2 \frac{\partial u}{\partial \tau} - \frac{1}{2} \frac{\partial (u^2)}{\partial x} \right\} = \frac{1}{12} \frac{\partial^4 u}{\partial x^4} \ , \tag{6.1.8}$$

and, provided that $u \to 0$ for $x \to \pm \infty$, to the KdV equation

$$u_\tau + \frac{1}{2} u u_x + \frac{1}{24} u_{xxx} = 0 \ . \tag{6.1.9}$$

Thus we see that the TL equation (6.1.1) includes the KdV equation (6.1.9) in the limit of $h \to 0_+$.

For example, the soliton solution

$$V_n(t) = \sinh^2 a \ \text{sech}^2 (a n \mp t \sinh a + \delta) \ , \tag{6.1.10}$$

of the TL equation (6.1.1) is transferred, by the transformations (6.1.2), (6.1.3) and by putting $a = \kappa h$, to the soliton solution

$$u(x, \tau) = \left\{ \frac{\sinh(\kappa h)}{h} \right\}^2 \text{sech}^2 \left[\kappa x + \left\{ \kappa \left(\frac{1}{h^2} - h^2 \right) \mp \frac{\sinh (\kappa h)}{\kappa h^3} \right\} \tau + \delta \right] \ , \tag{6.1.11}$$

of (6.1.7). If we take the limit of $h \to 0_+$, (6.1.11) reduces to

$$u(x,\tau) = \begin{cases} \kappa^2 \operatorname{sech}^2 \left(\kappa x - \dfrac{\kappa^3}{6} \tau + \delta \right) , & \text{(6.1.12a)} \\[4mm] \kappa^2 \operatorname{sech}^2 (\kappa x + \infty \tau + \delta) , & \text{(6.1.12b)} \end{cases}$$

according respectively to the upper and lower signs of (6.1.11). Equation (6.1.12a) is the soliton solution of the KdV equation (6.1.9). In (6.1.12b), ∞ means infinite velocity, so that even if there were a soliton at $\tau = 0$, it would vanish in the infinity $x = -\infty$ at the next instant ($\tau > 0$).

If we write the TL (6.1.1) by using *Hirota*'s bilinear form, we have [6.2]

$$[D_t^2 - 2(\cosh D_n - 1)] \, f_n \cdot f_n = 0 . \tag{6.1.13}$$

The transformations (6.1.2) and (6.1.4) lead to

$$D_t = h^3 D_\tau - (h - h^5) \, D_x , \tag{6.1.14a}$$

$$D_n = h D_x , \tag{6.1.14b}$$

and for small h, (6.1.13) gives

$$-2h^4 \left[D_x \left(D_\tau + \frac{1}{24} D_x^3 \right) f \cdot f \right] + O(h^6) = 0 , \tag{6.1.15}$$

which reduces, in the limit of $h \to 0_+$, to

$$D_x \left(D_\tau + \frac{1}{24} D_x^3 \right) f \cdot f = 0 . \tag{6.1.16}$$

This is nothing but the KdV equation (6.1.9) written in bilinear form.

Thus the KdV equation can be considered as a limit of the TL equation. Since the KdV is much easier to treat than the TL equation, it is convenient to see the rough idea of the behavior of the TL by looking at the solution of the corresponding case of the KdV. In such context, we sometimes refer to the KdV and its generalization.

6.2 Interacting Soliton Equations

Solitons deform when they collide, and recover their shape after they separate. We can keep the identity of each soliton during the process of encounter [6.3]. For example, we decompose the KdV equation as

$$\frac{\partial u_j}{\partial t} + 6u\frac{\partial u_j}{\partial x} + \frac{\partial^3 u_j}{\partial x^3} = 0 , \qquad (j = 1, 2, \ldots, N) , \tag{6.2.1a}$$

$$u = \sum_{j=1}^{N} u_j , \tag{6.2.1b}$$

where u_j stands for the jth soliton, and (6.2.1a) is the KdV equation for the N solitons. We can show that the area of u_j is conserved as

$$\int_{-\infty}^{\infty} u_j dx = 4\kappa_j . \tag{6.2.2}$$

In terms of the Lax pair

$$L = -\partial^2 - u , \tag{6.2.3a}$$

$$A = -4\partial^3 - 6u\partial - 3u_x , \tag{6.2.3b}$$

$(\partial = \partial/\partial x)$, the KdV equation (6.2.1) is written as

$$L\psi_j = -\kappa_j^2\psi_j , \tag{6.2.4a}$$

$$\frac{\partial}{\partial t}\psi_j = A\psi_j . \tag{6.2.4b}$$

κ_j and ψ_j being the eigenvalue and the eigenfunction.
Normalizing ψ_j as

$$\int_{-\infty}^{\infty} \psi_j^2 dx = 1 , \tag{6.2.5}$$

we have

$$u_j = 4\kappa_j\psi_j^2 . \tag{6.2.6}$$

For the TL equation, we may similarly decompose an N-soliton solution as

$$\frac{d}{dt}\left(\frac{dV_j(n)/dt}{1 + V(n)}\right) = V_j(n+1) + V_j(n-1) - 2V_j(n) , \qquad (j = 1, 2, \ldots, N) , \tag{6.2.7a}$$

$$V(n) = \sum_{j=1}^{N} V_j(n) . \tag{6.2.7b}$$

We can show that each soliton $V_j(n)$ keeps its identity in the course of interaction and reduces to the one-soliton solution when separated from other soli-

tons, or at $t \to \pm \infty$. For $V_j(n)$ of the interacting TL equation (6.2.7) we have properties similar to those expressed by (6.2.2 – 6), but they are much more intricate [6.3].

6.3 Integrability

Consider a classical conserved system described by f coordinates $q = (q_1, q_2, \ldots, q_f)$ and f momenta $p = (p_1, p_2, \ldots, p_f)$. If we know f independent first integrals $F_j(q_1, q_2, \ldots, q_f; p_1, p_2, \ldots, p_f)$ which are in involution, then we can solve the system by the method of quadrature. This is called the Liouville theorem for integrable systems.

The energy integral is one of the first integrals. Therefore systems with 1 degree of freedom ($f = 1$) are integrable. If a system with 2 degrees of freedom has conserved total momentum, or conserved total angular momentum, it too is integrable. But in general, systems with more than 2 degrees of freedom have insufficient first integrals, and so they are non-integrable (trivial exceptions are the linear systems).

As integrable systems with many degrees of freedom, we know the rigid body free to rotate around the fixed center of mass (*Euler*), and the axially symmetric top (Lagrange). A search for another integrable system was pursued by *Kovalevskaja*, who found an integrable asymmetric top [6.4].

The explicit solutions to Euler's and Lagrange's tops are expressed in terms of elliptic functions or derivatives of ϑ-functions, and therefore their singularities in the complex t-plane are poles, whose positions depend on the initial conditions. Kovalevskaja conjectured that the integrable system to be found would likewise have no movable singularities other than poles, and succeeded in finding the new integrable system.

The equations of motion for the periodic TL with N particles are written as $\dot{L} = AL - LA$ in terms of the Lax pair L and A, which are $N \times N$ matrices. We have N independent conserved quantities or integrals $\text{tr}\{L\}$, $\text{tr}\{L^2\}$, $\ldots$, $\text{tr}\{L^N\}$, and therefore the system is integrable in the sense of Liouville. Solutions to the TL equation are given in terms of the derivatives of ϑ-functions, with poles in the complex t-plane depending on initial conditions.

Though the conjecture that the systems with no movable singularities other than poles are integrable is not proved yet, it is widely accepted as quite efficient [6.5 – 7]. As an example of this short of Kovalevskaja's conjecture, let us apply it to the TL with two different masses.

Two-particle System. We consider a one-dimensional lattice of two particles with mass 1 and M respectively, and both ends fixed. The equations of motion are

$$\ddot{q}_1 = e^{-q_1} - e^{q_1 - q_2} \, , \tag{6.3.1a}$$

$$\ddot{q}_2 = e^{q_1 - q_2} - e^{q_2} \, . \tag{6.3.1b}$$

Putting [6.8]

$$z_1 = e^{q_1} , \qquad z_2 = e^{-q_2} , \tag{6.3.2}$$

we have

$$\ddot{z}_1 - \frac{z_1'^2}{z_1} = 1 - z_1^2 z_2 , \tag{6.3.3a}$$

$$\ddot{z}_2 - \frac{z_2'^2}{z_2} = 1 - z_1 z_2^2 . \tag{6.3.3b}$$

Assuming certain initial conditions, we assume that a singularity is located at t_0 (complex). In the vicinity of t_0 we let

$$z_1 \sim \frac{A}{(t-t_0)^\alpha} , \qquad z_2 \sim \frac{B}{(t-t_0)^\beta} . \tag{6.3.4}$$

Inserting (6.3.4) into (6.3.3) and comparing the dominant terms on both sides, we obtain

$$\alpha + \beta = 2 , \tag{6.3.5a}$$

$$\alpha = M\beta = -AB , \quad \text{so that} \tag{6.3.5b}$$

$$\alpha = \frac{2M}{1+M} , \quad \beta = \frac{2}{1+M} , \quad A = -\frac{\alpha}{B} . \tag{6.3.6}$$

If we change M from 0 to ∞, we see that α increases from 0 to 2, and β decreases from 2 to 0. We see that t_0 is a branch point if M is different from 0, 1, and ∞. If $M = 0$, 1 or ∞, then z_1 and z_2 have a pole at t_0. For $M = 0$ or ∞, the system is equivalent to a one-particle system, and of course integrable.

Therefore, only the case $M = 1$ ($\alpha = \beta = 1$) is to be examined to see if it actually leads to a general solution. We assume the Laurent expansion

$$z_1 = \frac{A}{t-t_0} \{ 1 + \gamma_1(t-t_0) + \gamma_2(t-t_0)^2 + \gamma_3(t-t_0)^3 + \ldots \} , \tag{6.3.7a}$$

$$z_2 = -\frac{1}{A(t-t_0)} \{ 1 + \delta_1(t-t_0) + \delta_2(t-t_0)^2 + \delta_3(t-t_0)^3 + \ldots \} , \tag{6.3.7b}$$

and insert them into (6.3.3). Then we see that including A and t_0, we have four arbitrary constants, namely A, t_0, γ_1, and γ_2. These are sufficient to satisfy the initial conditions of two positions and two velocities of the two particles in the lattice. We see $\delta_1 = -\gamma_1$, $\delta_2 = \gamma_2$ and other constants, higher than δ_3 and γ_3, are all determined by A, t_0, γ_1, and γ_2. Thus when $M = 1$, (6.3.7) provides the general solution, in accordance with the above conjecture.

6.4 Generalization of the TL Equation

One of the generalizations leads to a lattice of particles with the Hamiltonian

$$H = \frac{1}{2} \sum_{j=1}^{N} \frac{p_j^2}{m_j} + \sum_{k=1}^{M} \exp\left(\sum_{j=1}^{N} D_{kj} q_j\right) . \tag{6.4.1}$$

By a canonical transformation $(p_j/\sqrt{m_j} \to p_j, \; q_j \to q_j/\sqrt{m_j}$, we can rewite (6.4.1) as

$$H = \frac{1}{2} \sum_{j=1}^{N} p_j^2 + \sum_{k=1}^{M} \exp\left(\sum_{j=1}^{N} d_{kj} q_j\right) . \tag{6.4.2}$$

We may consider the further generalization,

$$H = \frac{1}{2} \sum_{i} \sum_{j} a_{ij} p_i p_j + \sum_{k} \exp\left(\sum_{j} d_{kj} q_j\right) . \tag{6.4.3}$$

The general Hamiltonian systems (6.4.2) or (6.4.3) are likely to have no integrals, except for the energy H. However, there are exceptional integrable systems including the usual TL. These integrable systems are obtained by assuming a representation in the Lax pair, that is, by assuming the equation of motion of the Lax form, $\dot{L} = AL - LA$. In this way, *Bogoyavlensky* [6.9] constructed integrable generalized TL by using the theory of simple Lie algebras. The relation between integrable systems and the Lie algebras has been extensively studied by *Olshanetsky* and *Perelomov*, *Kostant*, *Farwell* and *Minami*, and many other authors [6.10 – 12].
    Writing

$$K_N = \frac{1}{2} \sum_{j=1}^{N} p_j , \tag{6.4.4a}$$

$$V_N = \sum_{j=1}^{N} \exp\left(q_j - q_{j+1}\right) , \tag{6.4.4b}$$

we show some examples of these integrable systems:

$$H = K_{N+1} + V_N + \exp\left(q_{N+1} - q_1\right) \quad (N \geq 2) , \tag{6.4.5a}$$

$$H = K_N + \exp\left(-q_1 - q_2\right) + V_{N-1} + \exp\left(q_N\right) \quad (N \geq 2) , \tag{6.4.5b}$$

$$H = K_N + \exp(-2q_1) + V_{N-1} + \exp(2q_N) \quad (N \geq 3) , \tag{6.4.5c}$$

$$H = K_N + \exp(-q_1 - q_2) + V_{N-1} + \exp(q_{N-1} + q_N) \quad (N \geq 4) , \tag{6.4.5d}$$

$$H = K_7 + V_5 + \exp\left\{\frac{1}{2}(-q_1 - q_2 - q_3 + q_4 + q_5 + q_6) + \frac{1}{\sqrt{2}} q_7\right\}$$
$$+ \exp(-\sqrt{2}\, q_7) , \tag{6.4.5e}$$

$$H = K_8 + V_5 + \exp\left\{\frac{1}{2}(-q_1 + q_2 + \ldots + q_7 - q_8)\right\}$$
$$+ \exp(-q_1 - q_2) + \exp(-q_7 - q_8) , \tag{6.4.5f}$$

$$H = K_8 + V_5 + \exp\left\{\frac{1}{2}(-q_1 + q_2 + \ldots + q_7 - q_8)\right\}$$
$$+ \exp(-q_1 - q_2) + \exp(q_7 + q_8) , \tag{6.4.5g}$$

$$H = K_4 + \exp(-q_1 - q_4) + \exp(q_1 - q_2) + \exp(q_2 - q_3)$$
$$+ \exp(q_3) + \exp\left\{\frac{1}{2}(-q_1 - q_2 - q_3 + q_4)\right\} , \tag{6.4.5h}$$

$$H = K_3 + \exp(q_1 - q_2) + \exp(-2q_1 + q_2 + q_3)$$
$$+ \exp(q_1 + q_2 - 2q_3) . \tag{6.4.5i}$$

It is to be noted that these systems can be derived by using the above-mentioned Kovalevskaja's conjecture to find possibly integrable systems [6.13].

6.5 Two-Dimensional TL

Although many attempts have been made to generalize the TL equation to higher dimensions, exact treatment is so far restricted to systems discrete in one direction and continuous in other directions. *Mikhailov* [6.14] showed that the system described by

$$\frac{\partial^2 \phi_n}{\partial t^2} - \frac{\partial^2 \phi_n}{\partial x^2} = e^{\phi_{n+1} - \phi_n} - e^{\phi_n - \phi_{n-1}} , \tag{6.5.1}$$

is solvable by the inverse scattering method. This equation may be regarded as a discrete version of the partial differential equation

$$u_{tt} - u_{xx} - u_{yy} + h^2 u_{yyy} + h(u_y^2)_y = 0 , \tag{6.5.2}$$

or an extension of the Kadomtsev-Petviashvili equation (two-dimensional KdV equation)

$$(u_t + u u_x + u_{xxx})_x = u_{yy} \ . \tag{6.5.3}$$

If no x dependence is present, (6.5.1) reduces to the usual TL equation, and (6.5.2) to the Boussinesq equation.

The two-dimensional lattice equation (6.5.1) can be generalized [6.15] by replacing the r.h.s. of (6.5.1) by the generalized interaction of Bogoyavlensky's type, c.f. (6.4.2). To this generalized two-dimensional lattice one can apply the theory of simple Lie algebras to find out the cases where the equation can be written in Lax form [6.15].

The multi-dimensional TL of the Laplace type

$$a u_{xx}(x, y, n) + \beta u_{yy}(x, y, n)$$
$$= \exp \{u(x, y, n-1) - u(x, y, n)\} - \exp \{u(x, y, n) - u(x, y, n+1)\} \ , \tag{6.5.4}$$

was studied by *Nakamura* [6.16]. Introducing r by

$$r(x, y, n) = u(x, y, n) - u(x, y, n-1) \ , \tag{6.5.5}$$

and f by

$$\exp \{-r(x, y, n)\} - 1 = \left(a \frac{\partial^2}{\partial x^2} + \beta \frac{\partial^2}{\partial y^2} \right) \log f(x, y, n) \ , \tag{6.5.6}$$

we have

$$\exp \{-r(x, y, n)\} = \frac{f(n+1) f(n-1)}{f(n)^2} \ . \tag{6.5.7}$$

If we use Hirota's bilinear form, (6.5.4) is written as

$$[a D_x^2 + \beta D_y^2 - \{\exp (D_n) + \exp (-D_n) - 2\}] f(n) \cdot f(n) = 0 \ , \tag{6.5.8}$$

where D_x, for example, means

$$D_x^k a(x) \cdot b(x) = \partial_{x'}^k a(x+x') \, b(x-x')|_{x'=0} \ . \tag{6.5.9}$$

The soliton solution of (6.5.4) is

$$\exp \{u(n-1) - u(n)\} - 1 = \sinh^2 \frac{\omega}{2} \operatorname{sech}^2 [(kx + ly + \omega n)/2] \ , \tag{6.5.10}$$

with ω given by

$$\left[2 \sinh \frac{\omega}{2} \right]^2 = a k^2 + \beta l^2 \ . \tag{6.5.11}$$

Essentially (6.5.10) is a one-dimensional wave. We have a cylindrical solution of (6.5.4) of the form

$$f(x, y, n) = J_n(\rho) \ , \quad \text{with} \tag{6.5.12}$$

$$\rho = \left(\frac{x^2}{a} + \frac{y^2}{\beta}\right)^{1/2} \ , \tag{6.5.13}$$

where $J_n(\rho)$ stands for the Bessel function of degree n. Since $J_n(\rho)$ has zeros, $\exp(-r)$ given by (6.5.6) diverges. Equation (6.5.4) also has a particular solution

$$f(x, y, n) = 1 + \varepsilon^2 \sum_{m=n}^{\infty} J_m^2(\rho) \ , \tag{6.5.14}$$

where ε is an arbitrary constant. The wave associated with (6.5.14) is a "ripplon" rather than a soliton. More general solutions with the Bessel functions, *Bäcklund* transformation, etc. have been studied for such multi-dimensional TL equations [6.17]. By applying the inverse scattering method to (6.5.4), we can show that solitons and ripplons can be superposed nonlinearly.

6.6 Bethe Ansatz

The Bethe ansatz was first invented for quantum-mechanical one-dimensional spin-systems, and extended to one-dimensional systems of particles interacting via a repulsive delta-function potential. It was further developed by *C. N. Yang* and *C. P. Yang* [6.18, 19] to the thermodynamical system with a given temperature and pressure. *B. Sutherland* [6.20] was the first to apply the Bethe ansatz to the one-dimensional TL, the quantized TL. Since then, it has been conjectured that the Bethe ansatz will give exact results for the TL. Although the exactness has not been proven, at least in the classical limit it has been shown by *Opper* [6.21] that the Yang-Yang theory gives the exact free energy of the TL. We shall see this in the following.

In the classical limit, the Yang-Yang equations can be written as

$$\mu(\beta) = \frac{p^2}{2m} - \varepsilon(p) + \frac{1}{\beta} \int_{-\infty}^{\infty} \delta'(p-p') \, e^{-\beta\varepsilon(p')} \, dp' \ , \tag{6.6.1 a}$$

$$\beta P = \int_{-\infty}^{\infty} e^{-\beta\varepsilon(p)} dp \ , \tag{6.6.1 b}$$

where $\mu(\beta)$ is the chemical potential per particle at temperature $T = 1/k\beta$, and pressure P. p and m are the momentum and mass of the particle of the lattice, $\varepsilon(p)$ is related to a certain excitation energy, and $\delta(p)$ is the phase shift of collision with the relative momentum p. The chemical potential is given by the integral equation (6.6.1 a) under the subsidiary condition (6.6.1 b).

As for the TL, the interaction potential is $\phi(r) = (a/b)\,e^{-br} + ar + \text{const.}$ However, we regard the second term ar as a potential due to a pressure with the magnitude a. Then the pressure P in (6.6.1 b) is a plus external pressure ($P = a + P_{\text{ext}}$), and adjacent particles repel by the interaction potential

$$V(r) = A\,e^{-br} \quad (A = a/b) \ . \tag{6.6.2}$$

For such a potential the phase shift was calculated by *Mertens* [6.22]. In the classical limit we have

$$\delta(p) = -\frac{2p}{b}\left(\ln\frac{p}{\sqrt{mA}} - 1\right), \quad \text{so that} \tag{6.6.3}$$

$$\delta'(p) = -\frac{2}{b}\ln\frac{p}{\sqrt{mA}} \ . \tag{6.6.4}$$

For simplicity, we use the system of units where

$$m = 1, \quad A = 1, \quad b = 1 \ , \tag{6.6.5}$$

and put

$$x = \sqrt{\beta}\,p \ , \quad \phi(x) = e^{-\beta\varepsilon(p)}/\sqrt{\beta} \ . \tag{6.6.6}$$

Then (6.6.1 a) and (6.6.1 b) give

$$\beta\mu(\beta) - \frac{1}{2}\ln\beta - \beta P \ln\beta = \ln\phi(x) + \frac{x^2}{2}$$
$$- 2\int_{-\infty}^{\infty}\ln|x - x'|\,\phi(x')\,dx' \ , \tag{6.6.7a}$$

$$\beta P = \int_{\infty}^{-\infty}\phi(x)\,dx \ . \tag{6.6.7b}$$

We now have to solve these equations. Noting that the l.h.s. of (6.6.7 a) is independent of x, we differentiate the equation with respect to x, to give

$$\frac{d\phi}{dx} + x\phi = 2\phi\,T[\phi] \tag{6.6.8}$$

where T denotes the Hilbert transform given as

$$T[\phi] = P \int_{-\infty}^{\infty} \frac{\phi(x')}{x-x'} \, dx' \; , \tag{6.6.9}$$

(P denotes the principal value of the integral). Equation (6.6.8) is a nonlinear integro-differential equation with a singular kernel. We solve it following the method developed by *Satsuma* [6.23], under the condition (6.6.7b).

Here we merely state the results. We obtain

$$\phi(x) = \frac{1}{\pi} \, \mathrm{Im} \left\{ \frac{d}{dx} \ln f(x) \right\} , \quad \text{with} \tag{6.6.10}$$

$$f(x) = \int_{0}^{\infty} dt \; e^{\mathrm{i}xt} \, e^{-t^2/2} \, t^{\beta P - 1} \; . \tag{6.6.11}$$

Using some mathematical maneuvers, we can show that

$$\int_{-\infty}^{\infty} \ln|x'| \phi(x') \, dx' = -\ln \left\{ \frac{\Gamma(\beta P/2)}{\Gamma(\beta P)} \, 2^{\beta P/2 - 1} \right\} . \tag{6.6.12}$$

On the other hand, we put $x = 0$ in (6.6.7a) to give

$$\beta\mu = \ln \phi(0) - 2 \int_{-\infty}^{\infty} \ln|x'| \, \phi(x') \, dx' + \frac{1}{2} \ln \beta + \beta P \ln \beta \; . \tag{6.6.13}$$

With the help of the duplication formula for the Γ-function, we finally obtain

$$\beta\mu = -\ln \left\{ \sqrt{\frac{2\pi}{\beta}} \, \beta^{-\beta P} \, \Gamma(\beta P) \right\} . \tag{6.6.14}$$

If we follow the standard method of statistical mechanics, the chemical potential is given by the partition function for the lattice,

$$
\begin{aligned}
Z(\beta) &= e^{-\beta\mu} \\
&= \int_{-\infty}^{\infty} e^{-\beta p^2/2m} dp \int_{-\infty}^{\infty} \exp\left[-\beta \left\{ \frac{a}{b} \, e^{-br} + (a + P_{ext})r \right\} \right] dr \\
&= \frac{1}{b} \sqrt{\frac{2\pi m}{\beta}} \left(\frac{a}{b} \beta \right)^{-\beta P/b} \Gamma(\beta P/b) , \quad \text{with}
\end{aligned} \tag{6.6.15}
$$

$$P = a + P_{ext} \; . \tag{6.6.16}$$

In the simple system of units (6.6.5), we see that (6.6.15) coincides with (6.6.14). Thus it is shown that, in the classical case of the TL, the Bethe ansatz gives the correct thermodynamical result.

6.7 The Thermodynamic Limit

In the following, we consider the TL with the interaction potential

$$\phi(r) = \frac{a}{b}(e^{-br} - 1) + ar , \tag{6.7.1}$$

and, assuming that the lattice has a free end, we show [6.24] that the Hamiltonian of the lattice, with very large number of particles, is equivalent to

$$\mathcal{H} = \sum_{j=1}^{2N} \frac{p_j^2}{2m} + \sum_{r=1}^{N} \frac{ab}{2\pi} J_r + \sum_{r=1}^{N} n_r \varepsilon_r , \quad \varepsilon_r = 2\pi \tan \frac{\pi r}{2N} \tag{6.7.2}$$

where $2N$ is the number of particles, and

$$-\infty < p_j < \infty, \quad -\infty < J_r < \infty, \quad n_r = 0, 1, 2, \ldots , \tag{6.7.3}$$

specify the phase space.

We calculate the partition function $Z_{2N}^{(\mathcal{H})}(\beta)$ of the lattice using (6.7.2) and (6.7.3),

$$Z_{2N}^{(\mathcal{H})}(\beta) = \int_{-\infty}^{\infty} \ldots \int_{-\infty}^{\infty} dp_1 \ldots dp_{2N} \int_{0}^{\infty} \ldots \int_{0}^{\infty} dJ_1 \ldots dJ_N \sum_{n_1=0}^{\infty} \ldots \sum_{n_N=0}^{\infty} e^{-\beta \mathcal{H}}$$

$$= \left(\int_{-\infty}^{\infty} e^{-\beta p^2/2m} dp \right)^{2N} \left(\int_{0}^{\infty} e^{-\beta abJ/2\pi} dJ \right)^{N} \prod_{r=1}^{N} \sum_{n_r=0}^{\infty} e^{-\beta n_r \varepsilon_r}$$

$$= \left(\frac{2\pi m}{\beta} \right)^{N} \left(\frac{2\pi}{ab\beta} \right)^{N} \prod_{r=1}^{N} (1 - e^{-\beta \varepsilon_r})^{-1} . \tag{6.7.4}$$

We write $z = (a/b)\beta$, and obtain

$$\ln \prod_{r=1}^{N} (1 - e^{-\beta \varepsilon_r})^{-1} = -\sum_{r=1}^{N} \ln \left\{ 1 - \exp\left(-2\pi z \tan \frac{\pi r}{2N} \right) \right\}$$

$$= -\frac{2N}{\pi} \int_{0}^{\pi/2} d\theta \ln (1 - e^{-2\pi z \tan \theta}) , \tag{6.7.5}$$

in the limit of very large N. This is easily related to the so-called *Binet's* second formula [6.25] for the Γ-function,

$$\Gamma(z) = \sqrt{\frac{2\pi}{z}} \, z^z \, e^{-z} \exp\left\{ 2\int_{0}^{\infty} \frac{\tan^{-1}(t/z)}{e^{2\pi t} - 1} dt \right\} . \tag{6.7.6}$$

Integrating partially, we have

$$\Gamma(z) = \sqrt{\frac{2\pi}{z}}\, z^z\, e^{-z} \exp\left\{ -\frac{1}{\pi} \int_0^{\pi/2} d\theta \ln\left(1 - e^{-2\pi z\, \tan\,\theta}\right) \right\} . \tag{6.7.7}$$

Therefore, we obtain

$$Z_{2N}^{(\mathscr{H})}(\beta) = \left(\frac{2\pi m}{\beta}\right)^N \left[\frac{1}{b}\left(\frac{a}{b}\,\beta\right)^{-(a/b)\beta} e^{(a/b)\beta}\, \Gamma\left(\frac{a}{b}\,\beta\right) \right]^{2N} \tag{6.7.8}$$

On the other hand, the standard Hamiltonian of the same system is

$$H = \sum_{j=1}^{2N} \frac{p_j^2}{2m} + \sum_{n=1}^{2N} \phi(r_n) \quad (r_n = x_n - x_{n-1}) , \tag{6.7.9}$$

and the partition function is given by

$$Z_{2N}^{(H)}(\beta) = \int_{-\infty}^{\infty} \dots \int_{-\infty}^{\infty} dp_1 \dots dp_{2N} \int_{-\infty}^{\infty} \dots \int_{-\infty}^{\infty} dx_1 \dots dx_{2N}\, e^{-\beta H}$$

$$= \left(\int_{-\infty}^{\infty} e^{-\beta p^2/2m}\, dp \int_{-\infty}^{\infty} e^{-\beta \phi(r)}\, dr \right)^{2N}$$

$$= \left\{ \frac{1}{b} \sqrt{\frac{2\pi m}{\beta}} \left(\frac{a}{b}\,\beta\right)^{-(a/b)\beta} e^{(a/b)\beta}\, \Gamma\left(\frac{a}{b}\,\beta\right) \right\}^{2N} , \tag{6.7.10}$$

where we have used (6.6.15) putting $P_{\text{ext}} = 0$. Since (6.7.8) is the same as (6.7.10),

$$Z_{2N}^{(\mathscr{H})}(\beta) = Z_{2N}^{(H)}(\beta) , \tag{6.7.11}$$

we see that (6.7.2) is an effective Hamiltonian of the system. In (6.7.2) the potential energy of the lattice is split into certain collective parts; perhaps we may interpret the second term $abJ_r/2\pi$ as the potential energy of a ripple and the last term ε_r as the potential energy of a soliton. On the other hand, in (6.7.9), the energy is expressed as a sum of terms due to individual particles and springs. In the course of this calculation, it has been shown that

$$\ln \Gamma(z) = \ln \left(\sqrt{\frac{2\pi}{z}}\, z^z\, e^{-z} \right)$$

$$- \lim_{N \to \infty} \frac{1}{2N} \sum_{r=1}^{N} \ln \left\{ 1 - \exp\left(-2\pi z\, \tan\frac{\pi r}{2N} \right) \right\} . \tag{6.7.12}$$

6.8 Hierarchy of Nonlinear Equations

KdV Hierarchy. By fixing the operator

$$L = \partial^2 + 2u , \quad (\partial = \partial/\partial x) , \tag{6.8.1}$$

the Lax equation

$$\frac{\partial L}{\partial t} = AL - LA , \tag{6.8.2}$$

yields a system of nonlinear equations by assigning for A the operators A_0, $A_1, \ldots$, such that

$$A_0 = \partial \rightarrow u_t = u_x , \tag{6.8.3}$$

$$A_1 = \partial^3 + 3u\partial + \tfrac{3}{2}u_x \rightarrow u_t = 3uu_x + \tfrac{1}{4}u_{xxx} \rightarrow \text{(KdV equation)} . \tag{6.8.4}$$

Using

$$A_q = \partial^{2q+1} + \sum_{j=1}^{q} (b_j \partial^{2j-1} + \partial^{2j-1} b_j) \quad (q = 2, 3, \ldots) , \tag{6.8.5}$$

where the b_j's are certain functionals of u, u_x, $\ldots$, we can derive the higher order KdV equations. These constitute the KdV hierarchy.

KP Hierarchy. By using the properly defined operators ∂^{-1}, ∂^{-2}, $\ldots$, we let

$$L = \partial + u_2 \partial^{-1} + u_3 \partial^{-2} + \ldots , \tag{6.8.6}$$

and define B_n as the non-negative part of L^n. Then the Lax equation

$$\frac{\partial L}{\partial t_n} = B_n L - L B_n , \tag{6.8.7}$$

gives the Kadomtsev-Petviashvili (KP) hierarchy. For example, $u_2 = u$, $t_2 = y$, $t_3 = t$ gives

$$\frac{3}{4} u_{yy} = (u_t - 12uu_x - u_{xxx})_x , \tag{6.8.8}$$

which is called the KP equation, or the two-dimensional KdV equation. If the y dependence is discarded, the KP equation reduces to the KdV equation, and if we set $u_t = 0$, $y = t$, the KP equation gives the Boussinesq-type equation

$$3u_{tt} = (-12uu_x - u_{xxx})_x \ . \tag{6.8.9}$$

Thus the KP hierarchy includes the KdV, the Boussinesq-type and many other equations. The unified theory of solving the KP hierarchy has been developed by *Sato* and others [6.26], and this is considered one of the most outstanding achievements of the recent progress. Extending this idea Ueno developed the theory of the Toda hierarchy, which looks much more cumbersome than the KP case.

6.9 Some Numerical Results

The author started to think of the nonlinear lattice when tackling difficult problem of the energy transmission through a crystal lattice. The current theories of heat conduction assume Fourier's law, but this law has not yet been derived from first principles. The heat conductivity of a uniform linear lattice is infinite, and although impurities in the lattice hinder the energy flow they do not lead to Fourier's law. The behavior of such lattices has been numerically clarified and explained theoretically.

In general, localized oscillation takes place around a light impurity. When the concentration of impurities is large, nearly all the linear modes are localized, and energy flow becomes difficult because of this localization. It was conjectured that, if we introduce nonlinearity into the interaction between particles in the lattice, the waves would be mutually scattered and thus the nonlinearity would hinder the energy flow. However, computer experiments revealed that the strength of the energy flow in the nonlinear lattice was, in many cases, larger than that in the linear lattice with the same impurity concentration, that is, the nonlinearity actually enhanced the energy transmission. This is a typical example of a computer experiment that has demonstrated new facts different from the expectation of the common sense [6.28, 29].

According to our present knowledge, we understand that solitons carry energy in a nonlinear lattice. Solitons will not be influenced so much by the impurities and will give rise to higher conductivity compared with the linear case. In fact, before the discovery of the soliton, *Vissicher* et al.'s numerical computation [6.28] had revealed a remarkable soliton wave progressing faster than smaller ripples.

Localized Modes. Although we have a localized oscillation in a linear lattice around a light impurity, we may argue that this cannot happen in a nonlinear lattice. However, computer experiments on a nonlinear lattice seem to indicate that we can have localized oscillation around a light impurity. Figure 6.1 shows an example. This is the TL with a light impurity between the 49th and 50th nonlinear springs. A soliton is transmitted; as the soliton travels through the impurity, some energy is shared and after the passage of the soliton, the impurity continues oscillation.

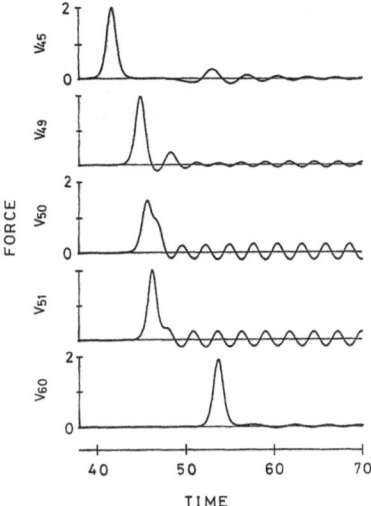

FORCE

V_{45} V_{49} V_{50} V_{51} V_{60}

40 50 60 70

TIME

Fig. 6.1. The localized oscillation around a light impurity with half the mass of the host particles

Energy Flow. As mentioned previously, the TL will become non-integrable when we introduce a mass impurity. It is known that the lattice with randomly introduced impurities sets up a constant temperature gradient except in the vicinity of the heat reservoir at both ends. According to a recent computer experiment [6.31], the energy flow in the regular diatomic TL with two kinds of masses distributed alternately is accompanied by a constant temperature gradient.

Nonlinear Breakdown. Since a soliton is a pulse with concentrated energy or strain, it may give rise to a breakdown if the material has a yielding point. This was tested numerically. At the same time, reflection of a soliton at a free end was also examined, since not so much was hitherto known of such reflection. When the parameters a and b of the potential $\phi(r) = (a/b)\,\mathrm{e}^{-br} + ar$ are positive, a soliton is a compressed pulse, and when it is reflected at a free end, it will change into rarefaction. Although a rarefactory pulse is unstable, and will thus change into some ripples, the instantaneous rarefactory pulse may give rise to a breakdown phenomenon if the lattice is weak against elongation. Such a lattice will be realized by the potential cut at $r = r_c$ (> 0), namely

$$\phi(r) = \begin{cases} \dfrac{a}{b}\,\mathrm{e}^{-br} + ar & (r < r_c)\ , \\[2mm] \dfrac{a}{b}\,\mathrm{e}^{-br_c} + ar_c & (r_c < r)\ . \end{cases} \qquad (6.9.1)$$

First, the lattice is stretched uniformly, giving a strain large enough but smaller than r_c, and then an end of the lattice is set free. A wave of contrac-

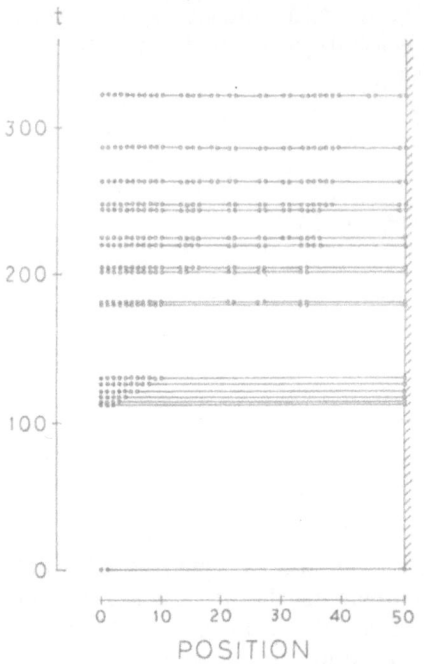

Fig. 6.2. An example of the chopping phenomena. A lattice of 50 particles is stretched uniformly, and cut at the left end. Successive breakdowns occur from the free end. The figure shows the state of the lattice at each instant when a new chopping occurs

tion will proceed to the fixed end, and when it is reflected there, many solitons will emerge. When the largest and the fastest soliton reaches the free end, it will give rise to a breakdown near the free end. Then it creates a new free end, and subsequent solitons will give rise to successive breakdown phenomena.

A result of the computer experiments [6.32] to verify the above assertion is given by Fig. 6.2. Some successive breakdowns (chopping phenomena) took place near the free end as expected. However, we see that breakdown happens not only at the free end, but also in other places. This is caused by the interference of waves, and each of these breakdown makes new free ends, which are seen to give rise to further breakdown.

Real solid materials, or large structures, might possibility experience such breakdown.

Appendices

Appendix A   Stieltjes' Method for a Continued Fraction [A.1]

Let $\gamma_1, \gamma_2, \ldots, \gamma_k$ be positive constants, and consider a series $y, y_1, y_2, \ldots$ defined by

$$\left.\begin{aligned}
y &= \gamma_1 y_1 + y_2 \\
y_1 &= \gamma_2 y_2 + y_3 \\
&\cdots\cdots\cdots \\
y_{N-1} &= \gamma_N y_N
\end{aligned}\right\}. \tag{A.1}$$

Rewriting we have

$$\left.\begin{aligned}
\frac{y}{y_1} &= \gamma_1 + \frac{y_2}{y_1} = \gamma_1 + 1 \bigg/ \frac{y_1}{y_2} \\
\frac{y_1}{y_2} &= \gamma_2 + \frac{y_3}{y_2} = \gamma_2 + 1 \bigg/ \frac{y_2}{y_3} \\
&\cdots\cdots\cdots \\
\frac{y_{N-1}}{y_N} &= \gamma_N
\end{aligned}\right\}. \tag{A.2}$$

Thus y/y_1 is expressed as a continued fraction

$$\frac{y}{y_1} = \gamma_1 + \cfrac{1}{\gamma_2 + \cfrac{1}{\gamma_3 + \cfrac{}{\ddots \\ + \cfrac{1}{\gamma_N}}}}. \tag{A.3}$$

On the other hand, by successive substitution, we have

$$\begin{aligned}
y &= (\gamma_1\gamma_2 + 1)\, y_2 + \gamma_1 y_3 \\
&= (\gamma_1\gamma_2\gamma_3 + \gamma_1 + \gamma_3)\, y_3 + (\gamma_1\gamma_2 + 1)\, y_4 \\
&= \cdots\cdots\cdots
\end{aligned} \tag{A.4}$$

Let us write

$$[\gamma_1] = \gamma_1, \qquad [\gamma_1, \gamma_2] = \gamma_1 \gamma_2 + 1 . \tag{A.5a}$$

and introduce the general notation

$$[\gamma_1, \gamma_2, \cdots, \gamma_k] = [\gamma_1, \gamma_2, \cdots, \gamma_{k-1}] \gamma_k + [\gamma_1, \gamma_2, \cdots, \gamma_{k-2}] \tag{A.5b}$$

with

$$[\gamma_l, \gamma_{l-1}] = 1 \tag{A.5c}$$

Then we have

$$y = [\gamma_1, \gamma_2, \cdots, \gamma_k] y_k + [\gamma_1, \gamma_2, \cdots, \gamma_{k-1}] y_{k+1} . \tag{A.6}$$

Multiplying γ_{k+1} on both sides, and using $\gamma_{k+1} y_{k+1} = y_k - y_{k+2}$ on the right-hand side, we have

$$[\gamma_{k+1}] y = [\gamma_1, \gamma_2, \cdots, \gamma_{k+1}] y_k - [\gamma_1, \gamma_2, \cdots, \gamma_{k-1}] y_{k+2} . \tag{A.7}$$

Further multiply γ_{k+2} and add (A.6); then the left-hand side gives $[\gamma_{k+1}, \gamma_{k+2}] y$. Repeating a similar procedure, we have the identity

$$[\gamma_{k+1}, \gamma_{k+2}, \cdots, \gamma_{k+l-1}] y = [\gamma_1, \gamma_2, \cdots, \gamma_{k+1}] y_k + (-1)^{l+1} [\gamma_1, \gamma_2, \cdots, \gamma_{k-1}] y_{k+l} . \tag{A.8}$$

Replacing k by l, and l by $m + n + 1$, we have

$$[\gamma_{l+1}, \cdots, \gamma_{l+m+n}] y = [\gamma_1, \cdots, \gamma_{l+m+n}] y_l + (-1)^{m+n} [\gamma_1, \cdots, \gamma_{l-1}] y_{l+m+n+1} . \tag{A.9}$$

On the other hand, eliminating y_{l+m+1} from

$$[\gamma_{l+1}, \cdots, \gamma_{l+m}] y = [\gamma_1, \cdots, \gamma_{l+m}] y_l + (-1)^m [\gamma_1, \cdots, \gamma_{l-1}] y_{l+m+1} \tag{A.10a}$$

$$[\gamma_{l+m+2}, \cdots, \gamma_{l+m+n}] y = [\gamma_1, \cdots, \gamma_{l+m+n}] y_{l+m+1} + (-1)^{n+1} [\gamma_1, \cdots, \gamma_{l+m}] y_{l+m+n+1} \tag{A.10b}$$

which can be derived from (A.8), we have a relation between y, y_l and $y_{l+m+n+1}$. Comparing this relation with (A.9), after some calculation, we have

$$[\gamma_{l+1}, \cdots, \gamma_{l+m}] [\gamma_1, \cdots, \gamma_{l+m+n}]$$
$$= [\gamma_1, \cdots, \gamma_{l+m}] [\gamma_{l+1}, \cdots, \gamma_{l+m+n}] + (-1)^m [\gamma_1, \cdots, \gamma_{l-1}] [\gamma_{l+m+2}, \cdots, \gamma_{l+m+n}] . \tag{A.11}$$

Especially for $l = n = 1$, we have

$$[\gamma_1, \cdots, \gamma_{m+2}] [\gamma_2, \cdots, \gamma_{m+1}] = [\gamma_1, \cdots, \gamma_{m+1}] [\gamma_2, \cdots, \gamma_{m+2}] + (-1)^m \qquad \text{(A.12)}$$

where we have used (A.5b).

Equations (A.11) and (A.10a) are of the same form, and if we put $m = 0$ in (A.11), it takes the same form as (A.6). Thus, when we replace γ_{2k+1} by $\gamma_{2k+1} z$, and define

$$\left.\begin{aligned}
Q_{2n}(z) &= [\gamma_1 z, \gamma_2, \gamma_3 z, \cdots, \gamma_{2n-1} z, \gamma_{2n}] \\
P_{2n}(z) &= [\gamma_2, \gamma_3 z, \gamma_4, \cdots, \gamma_{2n-1} z, \gamma_{2n}] \\
Q_{2n+1}(z) &= [\gamma_1 z, \gamma_2, \gamma_3 z, \cdots, \gamma_{2n}, \gamma_{2n+1} z] \\
P_{2n+1}(z) &= [\gamma_2, \gamma_3 z, \gamma_4, \cdots, \gamma_{2n}, \gamma_{2n+1} z]
\end{aligned}\right\} \qquad \text{(A.13)}$$

then (A.3) yields

$$\frac{Q_{2n}(z)}{P_{2n}(z)} = \gamma_1 z + \cfrac{1}{\gamma_2 + \cfrac{1}{\gamma_3 z + \cfrac{\ddots}{\qquad + \cfrac{1}{\gamma_{2n}}}}} \qquad \text{(A.14a)}$$

$$\frac{Q_{2n+1}(z)}{P_{2n+1}(z)} = \gamma_1 z + \cfrac{1}{\gamma_2 + \cfrac{1}{\gamma_3 z + \cfrac{\ddots}{\qquad + \cfrac{1}{\gamma_{2n} + \cfrac{1}{\gamma_{2n+1} z}}}}} \qquad \text{(A.14b)}$$

Now let

$$\beta_0 = \frac{1}{\gamma_1}, \quad \beta_n = \frac{1}{\gamma_n \gamma_{n+1}} \qquad (n \geq 1); \qquad \text{(A.15)}$$

then (A.14a) can be unified as

$$\frac{P_n(z)}{Q_n(z)} = \cfrac{\beta_0}{z + \cfrac{\beta_1}{1 + \cfrac{\beta_2}{z + \cfrac{\beta_3}{1 + \ddots}}}} \qquad \text{(A.16)}$$

where the last term at the bottom right is discarded. Further, noting the identity

$$z + \frac{\beta_1}{1 + \beta_2/z'} = z + \beta_1 - \frac{\beta_1\beta_2}{\beta_2 + z'} \tag{A.17}$$

with parameters z and z', we obtain

$$\frac{P_n(z)}{Q_n(z)} = \cfrac{\beta_0}{z + \beta_1 - \cfrac{\beta_1\beta_2}{z + \beta_2 + \beta_3 - \cfrac{\beta_3\beta_4}{z + \beta_4 + \beta_5 - \cdots}}} \tag{A.18}$$

Thus, if we write

$$\left.\begin{aligned}
\beta_0 = X_0, &\quad \beta_1 = Y_1 \\
\beta_1\beta_2 = X_1, &\quad \beta_2 + \beta_3 = Y_2 \\
\beta_3\beta_4 = X_2, &\quad \beta_4 + \beta_5 = Y_3 \\
\cdots\cdots &\quad \cdots\cdots
\end{aligned}\right\}, \tag{A.19}$$

we have the result

$$\frac{P_n(z)}{Q_n(z)} = \cfrac{X_0}{z + Y_1 - \cfrac{X_1}{z + Y_2 - \cfrac{X_2}{z + Y_3 - \cdots}}} \tag{A.20}$$

On the other hand, applying (A.11) to (A.10a) we have

$$Q_{2n}(z)\, P_{2n+1}(z) - P_{2n}(z)\, Q_{2n+1}(z) = 1 \tag{A.21}$$

or

$$\frac{P_{n+1}(z)}{Q_{n+1}(z)} - \frac{P_n(z)}{Q_n(z)} = \frac{(-1)^n}{Q_n(z)\, Q_{n+1}(z)}. \tag{A.22a}$$

If we put $n = 0$, we have formally

$$\frac{P_1(z)}{Q_1(z)} - \frac{P_0(z)}{Q_0(z)} = \frac{1}{Q_0(z)\, Q_1(z)} \tag{A.22b}$$

which is satisfied by defining $P_1(z) = Q_0(z) = 1$, and $P_0(z) = 0$. Then we have

$$\frac{P_n(z)}{Q_n(z)} = \frac{1}{Q_0(z)\, Q_1(z)} - \frac{1}{Q_1(z)\, Q_2(z)} + \cdots + \frac{(-1)^{n-1}}{Q_{n-1}(z)\, Q_n(z)}. \tag{A.23}$$

As can be seen from (A.5a), $P_n(z)$ and $Q_n(z)$ are polynomials of z, and if we write the highest terms,

$$\left.\begin{array}{l} P_{2n}(z) = \gamma_2\gamma_3 \cdots \gamma_{2n}z^{n-1} + \cdots \\[4pt] Q_{2n}(z) = \gamma_1\gamma_2 \cdots \gamma_{2n}z^n + \cdots \\[4pt] P_{2n+1}(z) = \gamma_2\gamma_3 \cdots \gamma_{2n+1}z^n + \cdots \\[4pt] Q_{2n+1}(z) = \gamma_1\gamma_2 \cdots \gamma_{2n+1}z^{n+1} + \cdots \end{array}\right\}. \tag{A.24}$$

Therefore $Q_n(z)Q_{n+1}(z)$ is a polynomial of the degree $n+1$, which may be written as

$$Q_n(z)\,Q_{n+1}(z) = cz(z + x_1)(z + x_2) \cdots (z + x_n) \tag{A.25}$$

with positive constants $c, x_1, x_2, \dots$. Thus we have

$$\frac{1}{Q_n(z)\,Q_{n+1}(z)} = \frac{\varepsilon_n^{(n)}}{z^n} - \frac{\varepsilon_{n+1}^{(n)}}{z^{n+1}} + \frac{\varepsilon_{n+2}^{(n)}}{z^{n+2}} - \cdots \tag{A.26}$$

where $\varepsilon_n^{(n)}, \varepsilon_{n+1}^{(n)}, \dots$ are some constants. We assume the limit

$$\lim_{n\to\infty} \frac{P_n(z)}{Q_n(z)} = \frac{P(z)}{Q(z)} \tag{A.27}$$

exists, and write

$$\frac{P(z)}{Q(z)} = \frac{c_0}{z} - \frac{c_1}{z^2} + \frac{c^2}{z^3} + \cdots - \frac{c_{2n-1}}{z^{2n}} + \frac{c_{2n}}{z^{2n+1}} - \cdots . \tag{A.28}$$

Then, from (A.23), we see, for example, that $P_{2n}(z)/Q_{2n}(z)$ has the same coefficients up to z^{-2n}, and can be written as

$$\frac{P_{2n}(z)}{Q_{2n}(z)} = \frac{c_0}{z} - \frac{c_1}{z_2} + \frac{c^2}{z^3} + \cdots - \frac{c_{2n-1}}{z^{2n}} + \frac{c'_{2n}}{z^{2n+1}} - \frac{c'_{2n+1}}{z^{2n+2}} + \cdots . \tag{A.29}$$

Since $Q_{2n}(z)$ is a polynomial of the degree n with respect to z, we may write

$$Q_{2n}(z) = \alpha_0 - \alpha_1 z + \alpha_2 z^2 + \cdots + (-1)^n\alpha_n z^n \tag{A.30}$$

and therefore

$$\begin{aligned} P_{2n}(z) - Q_{2n}(z)\frac{P(z)}{Q(z)} &= Q_{2n}(z)\left[\frac{P_{2n}(z)}{Q_{2n}(z)} - \frac{P(z)}{Q(z)}\right] \\[6pt] &= [(-1)^n z^n\alpha_n z^n + \cdots - \alpha_1 z + \alpha_0]\left(\frac{c'_{2n} - c_{2n}}{z^{2n+1}} - \frac{c'_{2n+1} - c_{2n+1}}{z^{2n+2}} + \cdots\right) \\[6pt] &= \frac{D_1}{z^{n+1}} + \frac{D_2}{z^{n+2}} - \cdots , \end{aligned} \tag{A.31}$$

where D_1, D_2, ... are constants. Thus, $Q_{2n}(z)P(z)/Q(z)$ coincides with $P_{2n}(z)$ except for terms higher than $z^{-(n+1)}$. However $P_{2n}(z)$ is a polynomial of z. Therefore, we see that the coefficients of $1/z$, $1/z^2$, ..., $1/z^n$ of $Q_{2n}(z)P(z)/Q(z)$, or of $Q_{2n}(-z) P(-z)/Q(-z)$, should vanish, namely

$$\alpha_0 c_k + \alpha_1 c_{k+1} + \cdots + \alpha_n c_{k+n} = 0 \tag{A.32}$$
$$(k = 0,1,2, \cdots, n - 1) .$$

We put $z = -\lambda$, and eliminate α_k from (A.30, 32) to obtain

$$Q_{2n}(-\lambda) = \frac{1}{\tilde{B}_n} \begin{vmatrix} 1 & \lambda & \lambda^2 & \cdots & \lambda^n \\ c_0 & c_1 & c_2 & \cdots & c_n \\ c_1 & c_2 & c_3 & \cdots & c_{n+1} \\ \multicolumn{5}{c}{\dotfill} \\ c_{n-1} & c_n & c_{n+1} & \cdots & c_{2n-1} \end{vmatrix} \tag{A.33}$$

where

$$\tilde{B}_n = \begin{vmatrix} c_1 & c_2 & \cdots & c_n \\ c_2 & c_3 & \cdots & c_{n+1} \\ \multicolumn{4}{c}{\dotfill} \\ c_n & c_{n+1} & \cdots & c_{2n-1} \end{vmatrix} . \tag{A.34}$$

Similarly, we have

$$Q_{2n+1}(-\lambda) = \frac{-1}{\tilde{A}_n} \begin{vmatrix} \lambda & \lambda^2 & \cdots & \lambda^{n+1} \\ c_1 & c_2 & \cdots & c_{n+1} \\ \multicolumn{4}{c}{\dotfill} \\ c_n & c_{n+1} & \cdots & c_{2n} \end{vmatrix} \tag{A.35}$$

with

$$\tilde{A}_n = \begin{vmatrix} c_0 & c_1 & \cdots & c_{n-1} \\ c_1 & c_2 & \cdots & c_n \\ \multicolumn{4}{c}{\dotfill} \\ c_{n-1} & c_n & \cdots & c_{2n-2} \end{vmatrix} . \tag{A.36}$$

By comparing with the coefficients of the highest terms in (A.24), we can prove that

$$\left.\begin{array}{l} \gamma_1\gamma_2 \cdots \gamma_{2n} = \dfrac{\tilde{A}_n}{\tilde{B}_n} \\[3mm] \gamma_1\gamma_2 \cdots \gamma_{2n+1} = \dfrac{\tilde{B}_n}{\tilde{A}_{n+1}} \end{array}\right\} \tag{A.37}$$

and thus we have

$$\gamma_{2n} = \frac{\tilde{A}_n^2}{\tilde{B}_n\tilde{B}_{n-1}}, \qquad \gamma_{2n+1} = \frac{\tilde{B}_n^2}{\tilde{A}_n\tilde{A}_{n+1}}. \tag{A.38}$$

Further, by using (A.15, 18) we obtain

$$X_n = \beta_{2n-1}\beta_{2n} = \frac{1}{\gamma_{2n-1}\gamma_{2n}^2\gamma_{2n+1}}. \tag{A.39}$$

Thus, the coefficients in the continued fraction (A.20) are expressed in terms of the coefficients of the partial fraction (A.28) as

$$X_n = \frac{\tilde{A}_{n-1}\tilde{A}_{n+1}}{\tilde{A}_n^2}. \tag{A.40}$$

These final results are derived without referring to coefficients higher than c_{2n}; they apply also to the finite continued fraction (A.20). For small values of n, we can verify that $\tilde{A}_{-1} = \tilde{A}_0 = 1$, $\tilde{B}_{-1} = 0$, and $\tilde{B}_1 = 1$ have consistent results.

For Y_n we have

$$\begin{aligned} Y_n &= \beta_{2n-2} + \beta_{2n-1} \\[2mm] &= \frac{1}{\gamma_{2n-2}\gamma_{2n-1}} + \frac{1}{\gamma_{2n-1}\gamma_{2n}} \\[2mm] &= \frac{\tilde{A}_n\tilde{B}_{n-2}}{\tilde{A}_{n-1}\tilde{B}_{n-1}} + \frac{\tilde{A}_{n-1}\tilde{B}_n}{\tilde{A}_n\tilde{B}_{n-1}}. \end{aligned} \tag{A.41}$$

If we put $z = -\lambda$, $X_0 = 1$, $X_n = a_{N-n}^2 (n \geq 1)$ and $Y_n = b_{N-n+1}$, then (A.20) gives $f(\lambda)$ of (3.10.36). On the other hand, if we write (3.10.22) in the form of a partial fraction (A.29), we can determine a_n^2 and b_n by using (A.40, 41).

We use notations $A_n \equiv (2a_n)^2$, $B_n \equiv 2b_n$, and $\varLambda \equiv -2z$ in (3.10.38) and below. Then, (3.10.51, 59, 62) give

$$\frac{\partial \varLambda_j}{\partial B_1} = \frac{C_j e^{-\varLambda_j t}/\prod^{(j)}(\varLambda_l - \varLambda_j)}{\sum_s C_s e^{-\varLambda_s t}\prod^{(s)}(\varLambda_l - \varLambda_s)} \equiv R_j \tag{A.42}$$

and from (3.10.52) we have

$$\frac{f_2(\varLambda)}{f_1(\varLambda)} = -\sum_{j=1}^{N} \frac{R_j}{\varLambda - \varLambda_j} = \sum_{k=0}^{\infty} (-1)^k \frac{c_k}{(-\varLambda)^{k+1}}, \tag{A.43}$$

where

$$c_k = (-1)^k \sum_{j=1}^{N} \Lambda_j^k R_j \ . \tag{A.44}$$

Equation (A.43) corresponds to (A.29), and (3.10.45) to (A.20), where $X_0 = 1$, $X_n = A_n$, and $Y_n = B_n$. Therefore $X_n = A_n = (2a_n)^2 = e^{-(Q_{n+1}-Q_n)}$ and $Y_n = B_n = 2b_n = \dot{Q}_n$ are given by (A.40, 41) as functions of c_k in (A.43, 44).

Appendix B   Suite de Sturm

Let

$$f_{n+1}(x) = (b_n x + c_n) f_n(x) + d_n f_{n-1}(x) \tag{B.1}$$

with

$$d_n > 0 \tag{B.2}$$

and

$$f_{-1}(x) = 0, \quad f_0(x) = 1 \tag{B.3}$$

(we may assume that $f_0 = \text{const.} > 0$). Some of b_n can be zero.

The roots of $f_n(x) = 0$ ($n = 1, 2, \ldots$) are called suite de Sturm. Now,

$$\left. \begin{aligned} f_1(x) &= b_0 x + c_0 \\ f_2(x) &= (b_1 x + c_1)(b_0 x + c_0) - d_1 \end{aligned} \right\} . \tag{B.4}$$

The root of $f_1(x) = 0$ ($x_1 = -c_0/b$) is between the two roots of $f_2(x) = 0$. This can be easily seen by using a graph (Fig. B.1). By mathematical induction, we can prove that the roots of $f_n(x) = 0$ are separated by the roots of $f_{n+1}(x) = 0$.

Appendix C   Simple Roots of $\Delta^2(\lambda) = 4$
Determine all the Roots [C.1]

Let λ_j^+ be the roots of $\Delta(\lambda) = 2$, and λ_j^- be those of $\Delta(\lambda) = -2$. Then we have

$$\Delta(\lambda) - 2 = A^{-1} \prod (\lambda - \lambda_j^+) = A^{-1} \prod (\lambda - \lambda_j^-) - 4 \tag{C.1}$$

($A = a_1 a_2 \ldots a_N$), which means that if λ_j^+'s are given, then they determine λ_j^-, and vice versa.

Let $\lambda_i^0 (i = 1, 2, \ldots, 2g + 2)$ be the simple roots of $\Delta^2(\lambda) = 4$, and λ_j be other roots. Then we may write

a

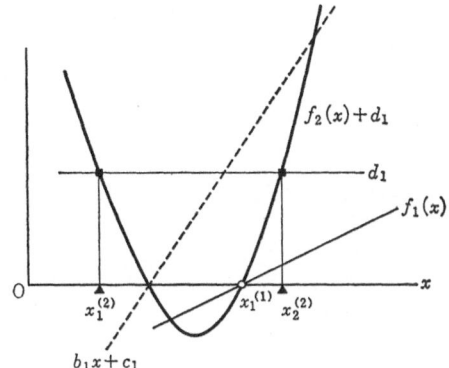

b

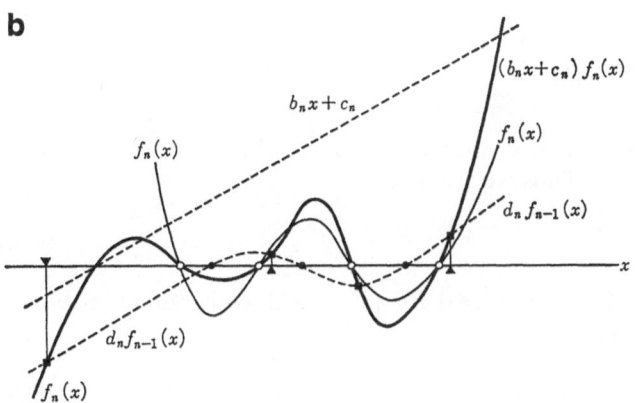

Fig. B.1. Illustration showing "suite de Strurm", $(\cdot)$ roots of $f_{n-1}(x) = 0$; $(\circ)$ $f_n(x) = 0$; $(\blacksquare - \blacktriangle)$ $f_{n+1}(x) = 0$

$$4 - \Delta^2(\lambda) = c_1 \prod_{i=1}^{2g+2} \left(1 - \frac{\lambda}{\lambda_i^0}\right) \prod \left(1 - \frac{\lambda}{\lambda_j}\right)^2 \tag{C.2}$$

with a certain constant c_1 (for brevity's sake, we assume that λ_i^0 and λ_j are different from zero).

On the other hand, let λ_j' be the root of $\Delta'(\lambda) = d\Delta(\lambda)/d\lambda = 0$ between λ_{2j}^0 and λ_{2j+1}^0. This is simple (we omit the proof). Since λ_j is a double root it is also a root of $\Delta'(\lambda) = 0$. Therefore we may write

$$\Delta'(\lambda) = c_2 \prod_{j=1}^{g} \left(1 - \frac{\lambda}{\lambda_j'}\right) \prod_k \left(1 - \frac{\lambda}{\lambda_k}\right) \tag{C.3}$$

with a certain constant c_2. Thus we have

$$\frac{\Delta'(\lambda)}{\sqrt{4 - \Delta^2(\lambda)}} = c_3 \frac{\prod(\lambda - \lambda_j')}{\sqrt{\prod(\lambda - \lambda_i^0)}} \tag{C.4}$$

where c_3 is a constant. Now let

$$\Delta(\lambda) = 2\cos\psi(\lambda) \; ;$$ (C.5)

then we have

$$\psi' = \frac{d\psi}{d\lambda} = -\frac{\Delta'(\lambda)}{2\sin\psi} = \pm\frac{\Delta'(\lambda)}{\sqrt{4-\Delta^2}}$$

$$= c_4 \frac{\prod(\lambda - \lambda_j')}{\sqrt{\prod(\lambda - \lambda_i^0)}}$$ (C.6)

with $c_4 = \pm c_3$. Integrating, we have

$$\psi(\lambda) = c_4 \int_{\lambda_1}^{\lambda} \frac{\prod_{j=1}^{g}(\lambda - \lambda_j')}{\sqrt{\prod_{i=1}^{2g+2}(\lambda - \lambda_i^0)}} d\lambda \, .$$ (C.7)

Therefore, for sufficiently large λ, we have $\psi(\lambda) \sim c_4 \ln\lambda + \ldots$. But then we have $\Delta(\lambda) \sim A^{-1}(\lambda^N + \ldots)$. Thus we see that

$$c_4 = \pm\sqrt{-1}\, N \, .$$ (C.8)

Further, since $\Delta^2 > 4$ and $\Delta' = $ real for $[\lambda_{2j-1}^0, \lambda_{2j}^0]$, ψ' is purely imaginary in this region and we have

$$\psi(\lambda_{2j-1}^0) - \psi(\lambda_{2j}^0) = \sqrt{-1}\, \varepsilon$$ (C.9)

where ε is a certain real number. However λ_{2j}^0 is a root of $\Delta^2 = 4$, which implies that $\cos\psi(\lambda_{2j}^0) = \pm 1$, or

$$\psi(\lambda_{2j}^0) = \pm m_j\pi$$ (C.10)

with some integer m_j. Therefore, we have

$$\pm 1 = \cos\psi(\lambda_{2j-1}^0) = \cos(\pm m_j\pi + \sqrt{-1}\,\varepsilon)$$

$$= \pm\cosh\varepsilon \, .$$ (C.11)

Thus, we have $\varepsilon = 0$, and

$$\psi(\lambda_{2j-1}^0) = (\psi\lambda_{2j}^0) \, .$$ (C.12)

We have therefore the simultaneous equations

$$\int_{\lambda_{2j-1}^0}^{\lambda_{2j}^0} \frac{\prod_{k=1}^{g}(\lambda - \lambda_k')}{\sqrt{\prod_{i=1}^{2g+2}(\lambda - \lambda_i^0)}} d\lambda = 0 \qquad (j = 1, 2, \cdots, g) \, .$$ (C.13)

These simultaneous equations determine λ_k' as a function of (λ_j^0), which means that the function $\psi(\lambda)$, and then all the roots of $\Delta^2(\lambda) = 4$ are determined by the simple roots λ_i^0.

Appendix D   All the Auxiliary Spectrum μ_j are Simple

The equation for $\varphi_1(n)$ is

$$a_n\varphi_1(n + 1) + b_n\varphi_1(n) + a_{n-1}\varphi_1(n - 1) = \lambda\varphi_1(n) . \tag{D.1}$$

Differentiating with respect to λ, and writing $\varphi_1' = d\varphi_1/d\lambda$, we have

$$a_n\varphi_1'(n + 1) + b_n\varphi_1'(n) + a_{n-1}\varphi_1'(n - 1) = \varphi_1(n) + \lambda\varphi_1'(n) . \tag{D.2}$$

Make the difference between the first equation multiplied by $\varphi_1'(n)$ and the second multiplied by $\varphi_1(n)$, and sum over n. Then we have

$$\sum_{n=1}^{N} [a_n\varphi_1(n + 1)\varphi_1'(n) - a_{n-1}\varphi_1(n)\varphi_1'(n - 1)]$$

$$- \sum_{n=1}^{N} [a_n\varphi_1(n)\varphi_1'(n + 1) - a_{n-1}\varphi_1(n - 1)\varphi_1'(n)]$$

$$= - \sum_{n=1}^{N} \varphi_1^2(n) . \tag{D.3}$$

Since the auxiliary spectrum μ_j satisfies

$$\left.\begin{array}{l} \varphi_1(N + 1, \mu_j) = 0 \\ \varphi_1'(0, \mu_j) = \varphi_1'(1, \mu_j) = 0 \end{array}\right\}, \tag{D.4}$$

we have

$$a_N\varphi_1(N, \mu_j)\,\varphi_1'(N + 1, \lambda)\Big|_{\lambda=\mu_j} = \sum_{n=1}^{N} \varphi_1^2(n, \mu_j) \neq 0 . \tag{D.5}$$

Therefore

$$\left.\begin{array}{l} \varphi_1(N, \mu_j) \neq 0 \\ \dfrac{d\varphi_1(N + 1, \lambda)}{d\lambda}\Big|_{\lambda=\mu_j} \neq 0 \end{array}\right\}. \tag{D.6}$$

The second equation implies that μ_j is simple.

Appendix E   **Lagrange's Interpolation Formula**

Consider a polynomial of the degree $n + 1$,

$$P(x) = \prod_{l=0}^{n} (x - x_l) \tag{E.1}$$

with constants x_l. By Cauchy's integral theorem, for a contour which encircles all the x_l, we have

$$\frac{1}{2\pi i} \oint \frac{z^m dz}{P(z)(z - x)} = \frac{x^m}{P(x)} + \sum_{j=0}^{n} \frac{x_j^m}{P'(x_j)(x_j - x)}. \tag{E.2}$$

On the other hand, this is the same as the integral over an infinitely large circle

$$\frac{1}{2\pi i} \oint \frac{z^m dz}{P(z)(z - x)} = \lim_{R \to \infty} \frac{R^m}{P(R)}. \tag{E.3}$$

Therefore we have

$$\frac{x^m}{P(x)} + \sum_{j=0}^{n} \frac{x_j^m}{P'(x_j)(x_j - x)} = \begin{cases} 0 & (0 \le m \le n) \\ 1 & (m = n + 1) \end{cases} \tag{E.4}$$

where

$$P'(x_j) = \prod_{l(\neq j)} (x_j - x_l). \tag{E.5}$$

If we put $x = 0$, we have

$$\sum_{j=0}^{n} \frac{x_j^{m-1}}{\prod\limits_{l(\neq j)0}^{n} (x_j - x_l)} = \begin{cases} 0 & (m < n + 1) \\ 1 & (m = n + 1) \end{cases} \tag{E.6}$$

or, changing the numbering

$$\sum_{j=1}^{g} \frac{x_j^s}{\prod\limits_{l(\neq j)=1}^{g} (x_j - x_l)} = \begin{cases} 0 & (s < g - 1) \\ 1 & (s = g - 1) \end{cases}. \tag{E.7}$$

For example, for $g = 2$,

$$\frac{1}{x_1 - x_2} + \frac{1}{x_2 - x_1} = 0$$

$$\frac{x_1}{x_1 - x_2} + \frac{x_2}{x_2 - x_1} = 1.$$

If $f(x)$ denotes a polynomial of the order n, then since

$$f(x) = \sum_{m=0}^{n} c_m x^m \tag{E.8}$$

we have

$$f(x) = \sum_{j=0}^{n} \frac{f(x_j) \, P(x)}{P'(x_j) \, (x - x_j)} \tag{E.9}$$

which is Lagrange's interpolation formula.

Appendix F   **Multivariable ϑ Function** (Riemann's ϑ Function)

This is defined by [F.1]

$$\vartheta(u_1, u_2, \ldots, u_g) = \sum_{m_1, m_2, \cdots, m_g=-\infty}^{m_1, m_2, \cdots, m_g=\infty} \exp\left(2\pi i \sum_{j=1}^{g} u_j m_j + \pi i \sum_{j,k=1}^{g} \tau_{jk} m_j m_k\right) \tag{F.1}$$

or by

$$\vartheta(u) = \vartheta(u\,|\,\tau) = \sum_{m} \exp(2\pi i u \cdot m + \pi i m \cdot \tau m) , \tag{F.2}$$

where

$$u = \begin{pmatrix} u_1 \\ u_2 \\ \vdots \\ u_g \end{pmatrix}, \quad m = \begin{pmatrix} m_1 \\ m_2 \\ \vdots \\ m_g \end{pmatrix}, \quad \tau = \begin{pmatrix} \tau_{11} & \cdots & \tau_{1g} \\ \tau_{21} & \cdots & \tau_{2g} \\ \cdots\cdots\cdots \\ \tau_{g1} & \cdots & \tau_{gg} \end{pmatrix} . \tag{F.3}$$

i) Let

$$e_k = \begin{pmatrix} 0 \\ \vdots \\ 1 \\ \vdots \\ 0 \end{pmatrix} \tag{F.4}$$

be a vector with the kth component, which is 1; then clearly

$$\vartheta(u + e_k\,|\,\tau) = \vartheta(u\,|\,\tau) . \tag{F.5}$$

Further, since

$$\vartheta(\boldsymbol{u}\,|\,\tau) = \sum_{\boldsymbol{m}} e^{m2\pi i(\boldsymbol{m}+\boldsymbol{e_k})\cdot\boldsymbol{u}+\pi i(\boldsymbol{m}+\boldsymbol{e_k})\cdot\tau(\boldsymbol{m}+\boldsymbol{e_k})}$$

$$= e^{2\pi i\boldsymbol{e_k}\cdot\boldsymbol{u}+\pi i\boldsymbol{e_k}\cdot\tau\boldsymbol{e_k}} \sum_{\boldsymbol{m}} e^{2\pi i\boldsymbol{m}\cdot\boldsymbol{u}+\pi i\boldsymbol{m}\cdot\tau\boldsymbol{m}+2\pi i\boldsymbol{m}\cdot\tau\boldsymbol{e_k}}$$

$$= e^{2\pi i\boldsymbol{e_k}\cdot\boldsymbol{u}+\pi i\boldsymbol{e_k}\cdot\tau\boldsymbol{e_k}} \sum_{\boldsymbol{m}} e^{2\pi i\boldsymbol{m}\cdot(\boldsymbol{u}+\tau\boldsymbol{e_k})+\pi i\boldsymbol{m}\cdot\tau\boldsymbol{m}}$$

$$\therefore \quad \vartheta(\boldsymbol{u}\,|\,\tau) = e^{2\pi i\boldsymbol{u_k}+\pi i\tau_{kk}}\,\vartheta(\boldsymbol{u}+\tau\boldsymbol{e_k}\,|\,\tau)\,, \tag{F.6}$$

if we write

$$\tau\boldsymbol{e_k} = \begin{pmatrix} \tau_{1k} \\ \tau_{2k} \\ \vdots \\ \tau_{gk} \end{pmatrix} = \tau_k \tag{F.7}$$

we have

$$\vartheta(\boldsymbol{u}+\tau_k\,|\,\tau) = e^{-2\pi i\boldsymbol{u_k}-\pi i\tau_{kk}}\,\vartheta(\boldsymbol{u}\,|\,\tau)\,. \tag{F.8}$$

ii) Let $\boldsymbol{x}$ be a vector,

$$\boldsymbol{x} = \begin{pmatrix} x_1 \\ x_2 \\ \vdots \\ x_g \end{pmatrix} \tag{F.9}$$

and $f(\boldsymbol{x})$ be a function of the form

$$f(\boldsymbol{x}) = \sum_{\boldsymbol{m}=-\infty}^{\infty} g(\boldsymbol{x}+\boldsymbol{m})\,. \tag{F.10}$$

Then $f(\boldsymbol{x})$ is a periodic function with period 1. Let its Fourier expansion be

$$f(\boldsymbol{x}) = \sum_{\boldsymbol{n}=-\infty}^{\infty} a_n e^{2\pi i\boldsymbol{n}\cdot\boldsymbol{x}} \tag{F.11}$$

where $\boldsymbol{n}$ is a integer vector; then

$$a_n = \int_0^1 f(\boldsymbol{y})\,e^{-2\pi i\boldsymbol{n}\cdot\boldsymbol{y}}\,d\boldsymbol{y}$$

$$= \sum_{\boldsymbol{m}=-\infty}^{\infty} \int_0^1 g(\boldsymbol{y}+\boldsymbol{m})\,e^{-2\pi i\boldsymbol{n}\cdot\boldsymbol{y}}\,d\boldsymbol{y}$$

$$= \int_{-\infty}^{\infty} g(\boldsymbol{y})\,e^{-2\pi i\boldsymbol{n}\cdot\boldsymbol{y}}\,d\boldsymbol{y}\,, \tag{F.12}$$

where $\boldsymbol{y}$ is a vector similar to $\boldsymbol{x}$. Therefore, we have

$$\sum_{m=-\infty}^{\infty} g(x + m) = \sum_{n=-\infty}^{\infty} e^{2\pi i n \cdot x} \int_{-\infty}^{\infty} g(y) e^{-2\pi i n \cdot y} \, dy \, . \tag{F.13}$$

If we put $x = 0$, we get

$$\sum_{m=-\infty}^{\infty} g(m) = \sum_{n=-\infty}^{\infty} \int_{-\infty}^{\infty} g(y) e^{-2\pi i n \cdot y} \, dy \tag{F.14}$$

which is Poisson's summation formula.

If we note that

$$\left. \begin{array}{l} \vartheta(u \mid \tau) = \displaystyle\sum_{m=-\infty}^{\infty} g(m) \\[2mm] g(m) = e^{2\pi i m \cdot u + \pi i m \cdot \tau m} \end{array} \right\} \tag{F.15}$$

by Poisson's summation formula, we obtain

$$\begin{aligned}
\vartheta(u \mid \tau) &= \sum_{n=-\infty}^{\infty} \int_{-\infty}^{\infty} e^{\pi i y \cdot \tau y + 2\pi i y \cdot u - 2\pi i n \cdot y} \, dy \\[2mm]
&= \sum_{n=-\infty}^{\infty} \int_{-\infty}^{\infty} e^{\pi i y \cdot \tau y + 2\pi i \, (u-n)} \, dy \\[2mm]
&= \sum_{n=-\infty}^{\infty} J(u - n) \, ,
\end{aligned} \tag{F.16}$$

where

$$J(x) = \int_{-\infty}^{\infty} e^{\pi i y \cdot \tau y + 2\pi i x y} \, dy \tag{F.17}$$

with $x = u - n$.

iii) To calculate $J(x)$, we introduce τ^{-1} which is the inverse of the matrix τ, and note that

$$(y + \tau^{-1}x) \cdot \tau(y + \tau^{-1}x) = y \cdot \tau y + 2xy + x \cdot \tau^{-1}x \, . \tag{F.18}$$

Then we have

$$\begin{aligned}
J(x) &= e^{-\pi i x \cdot \tau^{-1}x} \int_{-\infty}^{\infty} e^{\pi i (y + \tau^{-1}x) \cdot \tau(y + \tau^{-1}x)} \, dy \\[2mm]
&= e^{-\pi i x \tau^{-1}x} \int_{-\infty}^{\infty} e^{\pi i y \cdot \tau y} \, dy \, .
\end{aligned} \tag{F.19}$$

We assume that $i\tau$ is a symmetric matrix, introduce a unitary matrix S and its transpose S' to diagonalize $-\pi i\tau$, and let its eigenvalues be $a_l(l = 1, 2, \dots, g)$,

$$S'\tau S = -\frac{1}{\pi i}\begin{pmatrix} a_1 & & 0 \\ & \ddots & \\ 0 & & a_g \end{pmatrix}. \tag{F.20}$$

Then writing $y = Sz$, we have

$$\begin{aligned}
\int_{\infty-}^{\infty} e^{\pi i y \cdot \tau y}\, dy &= \int_{-\infty}^{\infty} e^{\pi i (Sz) \cdot \tau (Sz)}\, dz \\
&= \int_{-\infty}^{\infty} e^{\pi i z \cdot (S'\tau S) z}\, dz \\
&= \int_{-\infty}^{\infty} \cdots \int_{-\infty}^{\infty} e^{-(a_1 z_1^2 + a_2 z_2^2 + \cdots + a_g z_g^2)}\, dz_1 \cdots dz_g \\
&= \frac{\pi^{g/2}}{(a_1 a_2 \cdots a_g)^{1/2}} \\
&= \frac{\pi^{g/2}}{(-\pi i)^{g/2} |\tau|^{1/2}} \\
&= \frac{i^{g/2}}{|\tau|^{1/2}} \tag{F.21}
\end{aligned}$$

with

$$|\tau| = \det \tau. \tag{F.22}$$

Thus

$$\begin{aligned}
\vartheta(u \,|\, \tau) &= \frac{i^{g/2}}{|\tau|^{1/2}} \sum_{n=-\infty}^{\infty} e^{-\pi i (u-n) \cdot \tau^{-1} (u-n)} \\
&= \frac{i^{g/2}}{|\tau|^{1/2}} e^{-\pi i u \cdot \tau^{-1} u} \sum_{n=-\infty}^{\infty} e^{2\pi i n \cdot \tau^{-1} u - \pi i n \cdot \tau^{-1} n}. \tag{F.23}
\end{aligned}$$

Therefore

$$\vartheta(u \,|\, \tau) = \frac{i^{g/2}}{|\tau|^{1/2}} e^{-\pi i u \cdot \tau^{-1} u}\, \vartheta(\tau^{-1} u \,|\, -\tau^{-1}) \tag{F.24}$$

which is the imaginary transformation for a multivariable ϑ function. Especially for $g = 1$ we have Jacobi's imaginary transformation (4.7.16).

Appendix G   Hirota's Method

We write the equation of motion for an infinite lattice ($n = -\infty, \ldots, \infty$) in the form

$$m\frac{d^2 y_n}{dt^2} = a(e^{-br_{n-1}} - e^{-br_n})$$

$$r_n = y_{n+1} - y_n ,$$

and write

$$V_n(t) = a(e^{-br_n} - 1)$$

to get

$$\frac{d^2}{dt^2} \ln(1 + V_n) = V_{n+1} + V_{n-1} - 2V_n . \tag{G.1}$$

If we write the solution as

$$V_n(t) = \frac{d^2}{dt^2} \ln f_n(t) , \tag{G.2}$$

we have

$$\ddot{f}_n f_n - 2\dot{f}_n \dot{f}_n + f_n \ddot{f}_n - (f_{n+1} f_{n-1} - 2f_n f_n + f_{n-1} f_{n+1}) = 0 . \tag{G.3}$$

This can be rewritten in the form

$$\left[D_t^2 - 4 \sinh^2\left(\frac{D_n}{2}\right) \right] f_n \cdot f_n = 0 , \tag{G.4}$$

where D_t and D_n are defined by $(l = 1, 2, \ldots)$

$$\left. \begin{array}{l} D_t^l a(t) \cdot b(t) = \left(\dfrac{\partial}{\partial t} - \dfrac{\partial}{\partial t'} \right)^l a(t) b(t') \Big|_{t=t'} \\[4mm] D_n^l a(n) \cdot b(n) = \left(\dfrac{\partial}{\partial n} - \dfrac{\partial}{\partial n'} \right)^l a(n) b(n') \Big|_{n=n'} \end{array} \right\} . \tag{G.5}$$

Hirota (cf. [G.1]) began to use these operators to obtain multisoliton solutions and to discuss the Bäcklund transformation. Hirota's method is applied, not only to lattice systems, but also to the KdV equation, the sine-Gordon equation $[(\partial^2/\partial t^2 - \partial^2/\partial x^2) \varphi = \sin\varphi]$ and other systems.

Many identities are derived for the operators. For example, we have

$$\left. \begin{array}{l} D_x^m a \cdot 1 = \left(\dfrac{\partial}{\partial x} \right)^m a \\[4mm] D_x^m a \cdot b = (-1)^m D_x^m b \cdot a \end{array} \right\} . \tag{G.6}$$

If $F(D_t, D_n)$ is a polynomial of D_t and D_n, then we have in general

$$F(D_t, D_n)\, e^{\Omega_1 t + p_1 n} \cdot e^{\Omega_2 t + p_2 n}$$

$$= \frac{F(\Omega_1 - \Omega_2, p_1 - p_2)}{F(\Omega_1 + \Omega_2, p_1 + p_2)}\, F(D_t, D_n)\, \exp[(\Omega_1 + \Omega_2)\, t + (p_1 + p_2) n] \cdot 1\,. \quad \text{(G.7)}$$

Such identities are useful for calculation as we shall see in what follows.

We write the formal solution in the form (afterwards, we put $\varepsilon = 1$)

$$f_n(t) = 1 + \varepsilon f_n^{(1)} + \varepsilon^2 f_n^{(2)} + \cdots . \qquad \text{(G.8)}$$

From the coefficients of ε^1, we have

$$\left\{ \frac{\partial^2}{\partial t^2} - \left[2 \sinh\left(\frac{\partial}{2\partial n}\right) \right]^2 \right\} f_n^{(1)}(t) = 0 \qquad \text{(G.9a)}$$

and from those of ε^2,

$$2 \left\{ \frac{\partial^2}{\partial t^2} - \left[2 \sinh\left(\frac{\partial}{2\partial n}\right) \right]^2 \right\} f_n^{(2)}(t)$$

$$= - \left\{ D_t^2 - \left[2 \sinh\frac{D_n}{2} \right]^2 \right\} f_n^{(1)}(t) \cdot f_n^{(1)}(t)\,. \qquad \text{(G.9b)}$$

Higher order equations are also obtained. Solving these successively we have a multisoliton solution. If we assume a one-soliton solution in the form

$$f_n^{(1)}(t) = e^{2\eta}, \qquad f_n^{(2)}(t) = f_n^{(3)}(t) = 0\,, \qquad \text{(G.10a)}$$

(G.9) is satisfied by putting

$$\eta = \Omega t - pn - \eta_0, \qquad \Omega = \pm \sinh p \qquad \text{(G.10b)}$$

and (G.9b) is satisfied by $f_n^{(2)} = 0$.

To obtain a two-soliton solution, we assume

$$f_n^{(1)} = e^{2\eta_1} + e^{2\eta_2}\,. \qquad \text{(G.11)}$$

If we put

$$\eta_i = \Omega_i t - p_i n - \eta_i^0$$
$$\Omega_i = (\pm)_i \sinh p_i \qquad \text{(G.11')}$$

for $i = 1,2$ (η_i^0 are constants), then (G.9a) for ε^1 is satisfied. Since (G.7) gives

$$\left\{ D_t^2 - \left[2 \sinh\frac{D_n}{2} \right]^2 \right\} f_n^{(1)} \cdot f_n^{(1)}$$

$$= 2 \frac{[2(\Omega_1 - \Omega_2)]^2 - [2 \sinh(p_1 - p_2)]^2}{[2(\Omega_1 + \Omega_2)]^2 - [2 \sinh(p_1 + p_2)]^2} \left\{ \frac{\partial^2}{\partial t^2} - \left[2 \sinh\left(\frac{\partial}{2\partial n}\right) \right]^2 \right\} e^{2(\eta_1 + \eta_2)},$$

$$\text{(G.12)}$$

we see that (G.9b) is satisfied by

$$f_n^{(2)}(t) = e^{A+2\eta_1+2\eta_2} \tag{G.13a}$$

with

$$e^A = \frac{[2(\Omega_1 - \Omega_2)]^2 - [2\sinh(p_1 - p_2)]^2}{-[2(\Omega_1 + \Omega_2)]^2 + [2\sinh(p_1 + p_2)]^2} . \tag{G.13b}$$

In this case, it is easily shown that we can put $f_n^{(3)}(t) = f_n^{(4)}(t) = \dots = 0$. Thus we have a two-soliton solution.

We can discretize time variables by defining

$$\Delta_t F(t) = \delta^{-1}\left[F\left(t + \frac{1}{2}\delta\right) - F\left(t - \frac{1}{2}\delta\right)\right] \tag{G.14}$$

$$\Delta_t^2 F(t) = [F(t + \delta) + F(t - \delta) - 2F(t)]/\delta^2 \tag{G.15}$$

to obtain

$$\Delta_t^2 \ln(1 + V_n(t)) = \Delta_n^2 \delta^{-2} \ln(1 + \delta^2 V_n(t)) . \tag{G.16}$$

Hirota showed that this equation also gives multisoliton solutions.

If we put

$$V_n = \delta^{-2}\left\{\frac{f_n(t+\delta)f_n(t-\delta)}{f_n(t)^2} - 1\right\} \quad \text{or} \tag{G.17}$$

$$\delta^{-2} \ln(1 + \delta^2 V_n) = \Delta_t^2 \ln f_n(t) \tag{G.18}$$

(G.16) reduces to

$$\delta^{-2}\{f_n(t+\delta)f_n(t-\delta) - f_n(t)^2\} = f_{n+1}(t)f_{n-1}(t) - f_n(t)^2 \tag{G.19}$$

which is the discrete-time TL equation.

Perk [G.2] found that the correlation function at the critical point of the two dimensional Ising model satisfies the discrete (imaginary) time TL equation.

Hirota [G.3] generalized the TL equation and obtained the three dimensional difference equation,

$$\alpha f(\lambda+1, \mu, \nu)f(\lambda-1, \mu, \nu) + \beta f(\lambda, \mu+1, \nu)f(\lambda, \mu-1, \nu)$$
$$+ \gamma f(\lambda, \mu, \nu+1)f(\lambda, \mu, \nu-1) = 0 \tag{G.20}$$

where α, β and γ are arbitrary parameters. In some appropriate limits of parameters, (G.20) gives various soliton equations including the TL, the discrete-time TL, the two dimensional TL, the KdV, the KP, tine sine-Gordon and the Benjamin-Ono equations. When $\alpha + \beta + \gamma = 0$, (G.20) gives the KP-hierarchy.

Appendix H   Induction Phenomenon

If we excite a vibrational mode of a lattice, energy will be transformed to other modes through nonlinear interaction.

We consider a one-dimensional lattice with both ends ($n = 0$, and $n = N$) fixed. We assume quadratic nonlinear interaction potential (which is more convenient for numerical calculation). Then the Hamiltonian is

$$\left. \begin{aligned} \mathcal{H} &= \sum_{j=1}^{N}\left[\frac{1}{2}\,\dot{x}_j^2 + \frac{1}{2}\,(x_j - x_{j-1})^2 + \frac{\lambda}{4}\,(x_j - x_{j-1})^4\right] \\ x_0 &= x_N = 0 \end{aligned} \right\} \tag{H.1}$$

We introduce normal coordinates a_k by

$$x_j = \mathrm{i}\sum_{k=-N}^{N}\frac{a_k}{\omega_k}\,\mathrm{e}^{-\mathrm{i}\pi jk/N} \tag{H.2}$$

where ω_k is the eigenfrequency of the linear mode,

$$\omega_k = 2\sin(\pi k/2N) \qquad (-N \le k \le N), \tag{H.3}$$

and

$$a_0 = 0 \tag{H.4a}$$
$$a_k = a_{-k}\,. \tag{H.4b}$$

The equations of motion are

$$\ddot{a}_k = -\omega_k^2 a_k - \lambda\omega_k^2\sum_{k'}\sum_{k''}\sum_{k'''}a_{k'}a_{k''}a_{k'''}D(k + k' + k'' + k''') \tag{H.5}$$
$$(k = -N, \cdots, N)$$

with

$$\omega_k^2 D(k + k' + k'' + k''') = -\omega_k\sin\frac{\pi(k' + k'' + k''')}{2N}\frac{1}{2N + 1}$$
$$\sum_{j=-N}^{N}\mathrm{e}^{\mathrm{i}\pi j(k+k'+k''+k''')} \tag{H.6}$$

The maximum value of $|k + k' + k'' + k'''|$ is $4N$, and we have

$$k + k' + k'' + k''' = 0, \qquad D(0) = 1 \tag{H.7a}$$
$$k + k' + k'' + k''' = \pm 2N, \qquad D(\pm 2N) = -1 \tag{H.7b}$$
$$k + k' + k'' + k''' = \pm 4N, \qquad D(\pm 4N) = 1\,. \tag{H.7c}$$

For other values of $k + k' + k'' + k'''$, we have $D(k + k' + k'' + k''') = 0$. Equation (H.7a) gives selection rules [H.1].

If we excite mode k_0, and keep $a_k = 0$ for an other mode initially,

$$\ddot{a}_{k_0}a = -\omega_{k_0}^2 a_{k_0} - 3\lambda\omega_{k_0}^2 a_{k_0}^3 \tag{H.8}$$

which gives initial oscillation of a_{k_0} in terms of an elliptic function.

In general, we have the following sets of k', k'' and k''' which satisfy $k + k' + k'' + k'''$ in (H.7a):

$$k' = -k, \quad k'' = -k''' = k_0 \qquad \text{and the equivalent (6 sets)} \tag{H.9}$$

$$-k' = k'' = -k''' = k \qquad \text{and the equivalent (3 sets).} \tag{H.10}$$

Therefore (H.5) gives

$$\ddot{a}_k = -\omega_k^2 a_k - 6\lambda\omega_k^2 a_{k_0}^2 a_k - 3\lambda\omega_k^2 a_k^3 - \text{(other terms)} . \tag{H.11}$$

The second term of (H.11) comes from (H.9) and the third from (H.10).

If no other terms satisfy $D(k + k' + k'' + k''') \neq 0$, and if initially $a_k = 0$, then $a_k = 0$ all the time. This means that mode k is not excited by k_0.

However, for the special value such that

$$k_1 = 3k_0 \tag{H.12}$$

or

$$k_1 = \pm(2N - 3k_0) , \tag{H.13}$$

the set of values

$$|k'| = |k''| = k''' = k_0 \tag{H.14}$$

satisfies (H.7a) or (H.7b),

$$D(k_1 + k' + k'' + k''') \neq 0$$

Therefore, for such values of k_0, we have the equation of motion

$$\ddot{a}_{k_1} + \omega_{k_1}^2(1 + 6\lambda a_{k_0}^2) a_k + 3\lambda\omega_{k_1}^2 a_{k_1}^3 = \pm\lambda\omega_{k_1}^2 a_{k_0}^2 \tag{H.15}$$

where $\pm$ signs on the right-hand side are due to (H.12, 13). Since the last term on the left-hand side is proportional to $a_{k_1}^3$ and stays small compared with other terms, we neglect it to have

$$\ddot{a}_{k_1} + \omega_{k_1}^2(1 + 6\lambda a_{k_0}^2) a_k = \pm\lambda\omega_{k_1}^2 a_{k_0}^2 . \tag{H.16}$$

The term $6\lambda a_{k_0}^2$ on the left-hand side is due to (H.9), or $k - k + k_0 - k_0 = 0$ in this case.

For example, let us assume

$$N = 16, \qquad k_0 = 11 .$$ (H.17)

We excite the 11th mode; then (H.13) gives

$$k_1 = 3k_0 - 2N = 1 .$$ (H.18)

Therefore

$$k_1 = 1$$ (H.19)

is excited. Since $D(2N) = -1$ in this case, we have

$$\ddot{a}_1 + \omega_1^2(1 + 6\lambda a_{11}^2) \, a_1 = + \lambda \omega_1^2 a_{11}^3 .$$ (H.20)

After a_1 is thus excited, $k = 3$ is excited through $k_3 - k_1 - k_1 - k_1 = 0$. That is, (H.9) is satisfied not only by $3 - 3 + 11 - 11 = 0$, but also by

$$3 - 1 - 1 - 1 = 0$$

and this coupling excites $k = 3$, leading to the equation for the excitation of $k = 3$ through a_1,

$$\ddot{a}_3 + \omega_3^2(1 + 6\lambda a_{11}^2) \, a_3 = - \lambda \omega_3^2 a_1^3 .$$ (H.21)

After $k_1 = 1$ is excited, we have another energy flow because (H.9) is satisfied, besides $9 - 9 + 11 - 11 = 0$, by

$$9 - 11 + 1 + 1 = 0$$

thus leading to excitation of $k = 9$ through a_1,

$$\ddot{a}_9 + \omega_9^2(1 + 6\lambda a_{11}^2) \, a_9 = - 3\lambda \omega_9^2 a_1^2 a_{11} .$$ (H.22)

Similarly, since (H.9) is satisfied, besides $13 - 13 + 11 - 11 = 0$, by

$$13 - 11 - 1 - 1 = 0 ,$$

the mode $k = 13$ is excited through a_1,

$$\ddot{a}_{13} + \omega_{13}^2(1 + 6\lambda a_{11}^2) \, a_{13} = - 3\lambda \omega_{13}^2 a_1^2 a_{11} .$$ (H.23)

In addition, for $k = 13$, we have

$$13 - 9 - 3 - 1 = 0 .$$

Therefore $k = 13$ is also excited through $k = 9$, $k = 3$, and $k = 1$.

Thus, if we initially excite the mode $k_0 = 11$ of a lattice with $N = 16$, firstly the mode $k = 1$ is excited, and then $k = 3$, 9, and 13 are successively excited. Energy will flow further to other modes. However, energy of these modes may not increase. Energy of a mode can increase only if it is in an unstable region. That is, energy of a mode can be excited to have greater energy only when it satisfies the following two conditions:

I) selection rules,

II) instability conditions.

If we take the example, $n = 16$, $k_0 = 11$, the initial state is

$$a_{11} = a \cos \omega_{11} t .\tag{H.24}$$

This state continues for a while, and (H.20–23) are the Mathieu equations

$$\frac{d^2 a_k}{d\tau^2} + (\alpha_k + 2h_k^2 \cos 2\tau) a_k = 0 \quad \text{with}\tag{H.25}$$

$$\left.\begin{aligned}
\tau &= \omega_{11} t \\
\alpha_k &= \left(\frac{\omega_k}{\omega_{11}}\right)^2 (1 + 3\lambda a^2) \\
h_k^2 &= \left(\frac{\omega_k}{\omega_{11}}\right)^2 \frac{3\lambda a^2}{2}
\end{aligned}\right\}.\tag{H.26}$$

α_k and h_k^2 are determined as functions of λa^2 and ω_k from (H.26), which may be conveniently given as the crossing point of two lines

$$\alpha_k = \left(2 + \frac{2}{3\lambda a^2}\right) h_k^2\tag{H.27}$$

$$\alpha_k = 2h_k^2 + \left(\frac{\omega_k}{\omega_{11}}\right)^2 ,\tag{H.28}$$

where (H.27) is determined when λa^2 is given, and the line (H.28) performs a parallel shift when k, or ω_k, is changed.

For example, the case where

$$\lambda a^2 = 0.195\tag{H.29}$$

is shown in Fig. H.1. In this figure, unstable regions of the Mathieu equation are also indicated. When $\lambda a^2 = 0.195$, we see that only $k = 9$ is in an unstable region. Therefore $k = 9$ is the only mode whose energy will increase. Figure H.2 shows the result of numerical experiments for this case [H.1]. The energy of each mode is shown in Figs. H.3 and H.4. Firstly the mode $k = 1$ is excited, but

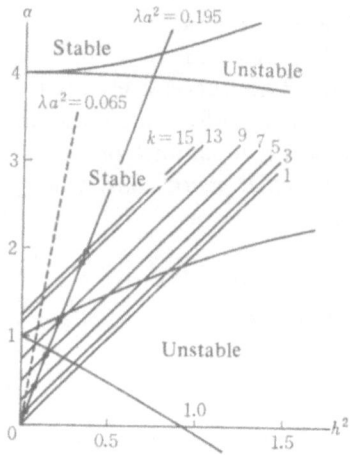

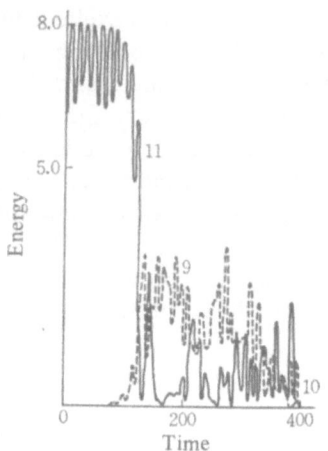

Fig. H.1. Stable and unstable regions (*Saito* et al. [H.1])

Fig. H.2. An example of induction phenomena (*Saito* et al. [H.1])

its energy never increases much, and after a while, the mode $k = 9$ and then $k = 13$ obtain greater energy. Further, by similar cascade processes, other modes begin to have greater energy, and after some time, the initially excited mode (in this case $k_0 = 11$) loses energy rather abruptly. This is called the induction phenomenon. As is seen from the above example, when the nonlinearity is of the fourth order, if we initially excite the odd (even) mode, then even (odd) modes will never be excited. The above selection rules prohibit energy transfer between modes with different parity. However, due to inevitable errors of calculation and errors of the initial data, a mode with parity different from the intended initial exitation may be excited, and then, through this mode, energy may flow to all the modes. The fact that modes $k = 10$ and 2 obtain energy after a long time is to be interpreted in this manner.

Thus, when some modes are excited, energy flows through these modes. If, besides k_0, we initially give a small amount of energy to some other mode (seeding), then the induction time is diminished. This is a well-known fact of computer experiments [H.2].

If the nonlinear parameter λ is large, and the energy given to the system is sufficiently large, then energy partition will be achieved as far as the selection rules admit [H.3]. Even for the lattice with exponential interaction, such behavior is known [H.4]. It is also to be noticed that when the size of the system is increased, unstable regions come closer together and, consequently, energy partition is enhanced.

On the contrary, if the nonlinear parameter is small and the energy of the system is too small, unstable regions will be ruled out, energy partition will not occur and the system will exhibit periodic behavior. Since the FPU computer

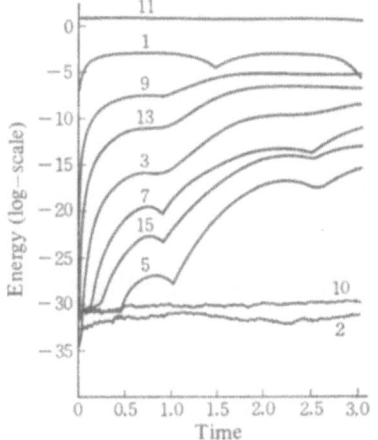

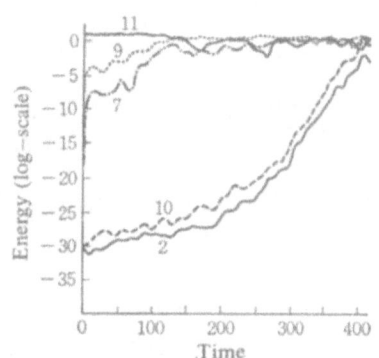

Fig. H.3. Energy sharing—short time (*Saito* et al. [H.1])

Fig. H.4. Energy sharing—long time (*Saito* et al. [H.1])

experiment did not satisfy the instability condition, it exhibited periodic recurrence phenomena.

Energy partition does not mean ergodicity. Even in a linear lattice, energy partition with respect to particle energy is generally achieved in a short time. In a nonlinear lattice energy partition among linear modes is also easily accomplished.

Appendix I   Partition Function

For a lattice with exponential interaction, the classical partition function can be rigorously calculated [I.1].

The partition function for an ensemble with constant pressure is given, for a one-dimensional system, as [I.2]

$$Z(p, T) = \left(\frac{2\pi mkT}{h^2}\right)^{N/2} Q^N \tag{I.1}$$

with

$$Q = \int_{-\infty}^{\infty} e^{-[\phi(r)+pr]/kT}\, dr \ . \tag{I.2}$$

For a lattice with exponential interaction,

$$Q = \int_{-\infty}^{\infty} \exp\left\{-\left[\frac{a}{b}\,(e^{-br}-1) + (a+p)r\right]\Big/ kT\right\} dr \ . \tag{I.3}$$

If we put

$$y = e^{-br},$$ (I.4)

we have

$$Q = \frac{1}{b} e^{a/bkT} \int_0^\infty y^{(a+p)/bkT-1} e^{-ay/bkT} \, dy$$

$$= \frac{e^{a/bkT}}{b} \left(\frac{bkT}{a}\right)^{(a+p)/bkT} \Gamma\left(\frac{a+p}{bkT}\right)$$ (I.5)

when $\Gamma(x)$ denotes the Γ function. The thermal expansion of the lattice is given as the average of r,

$$\bar{r} = \frac{\int_{-\infty}^\infty r e^{-[\phi(r)+pr]/kT} \, dr}{\int_{-\infty}^\infty e^{-[\phi(r)+pr]/kT} \, dr}$$

$$= -kT \frac{\partial}{\partial p} \ln Q$$

$$= -\frac{1}{b}\left[\ln \frac{bkT}{a} + \psi\left(\frac{a+p}{bkT}\right)\right]$$ (I.6)

and the average energy (enthalpy) per particle is given as

$$\frac{E}{N} = \frac{1}{2} kT + \bar{\phi}(r)$$

$$\bar{\phi}(r) = -\frac{\partial}{\partial(1/kT)} \ln Q = (a+p)\bar{r} + \frac{p}{b}$$ (I.7)

where $\bar{\phi}(r)$ denotes the average of the potential energy. In (I.6), ψ means the digamma function

$$\psi(x) = \frac{d \ln \Gamma(x)}{dx}.$$ (I.8)

For low temperature, or for small nonlinearity, we have

$$\bar{r} = \frac{kT}{2(a+p)}.$$ (I.9)

Appendix J   Dispersion Relation for the Lattice Without Expansion

I) A cnoidal wave solution (2.3.1) is

$$e^{-br_*} - 1 = \frac{(2K\nu)^2}{ab/m} \left\{ dn^2 \left[2 \left(\frac{n}{\lambda} \mp \nu t \right) K \right] - \frac{E}{K} \right\}. \tag{J.1}$$

Because of the formula

$$\nu \int_0^{1/\nu} dn^2 \left[2 \left(\frac{n}{\lambda} \mp \nu t \right) K \right] dt = \frac{E}{K}, \tag{J.2}$$

the time average gives

$$\overline{e^{-br_*} - 1} = 0. \tag{J.3}$$

This means that the average of the force $f(r) = -d\phi/dr$ for the exponential interaction $\phi(r) = (a/b) e^{-br} + ar$ vanishes;

$$\overline{f(r)} = a(\overline{e^{-br_*} - 1}) = 0. \tag{J.4}$$

Therefore (J.1) is a solution when the average of the force of the nonlinear spring vanishes, or when there is no external force on the lattice.

Even if we apply a constant pressure on the lattice, this force does not appear in the equations of motion. However, we can modify the interaction potential in such a way that it includes the external force. Suppose we have a constant external pressure f_{ext} applied to the lattice. This is equivalent to saying that each spring has additional potential energy $f_{\text{ext}}r$, so that the effective potential is

$$\phi_{\text{eff}}(r) = \frac{a}{b} e^{-br} + (a + f_{\text{ext}}) r = \frac{a'}{b} e^{-b(r+\sigma')} + a'r, \tag{J.5}$$

where we have introduced a constant σ' by

$$a' \equiv a + f_{\text{ext}} = ae^{b\sigma'}. \tag{J.6}$$

Since $\phi_{\text{eff}}(r)$ has a minimum at $r = -\sigma'$, we see that σ' is the contraction of a spring due to a static pressure f_{ext}.

Thus we see that the equations of motion (2.3.9) have, besides (J.1), a solution

$$e^{-b(r_*+\sigma')} - 1 = \frac{(2K\nu)^2}{a'b/m} \left\{ dn^2 \left[2 \left(\frac{n}{\lambda} \mp \nu t \right) K \right] - \frac{E}{K} \right\} \tag{J.7}$$

where the dispersion relation is given from (2.3.2) by replacing a by a', or

$$2K\nu = \sqrt{\frac{a'b}{m}} \Big/ \sqrt{\frac{1}{\mathrm{sn}^2(2K/\lambda)} - 1 + \frac{E}{K}} \ . \tag{J.8}$$

The force of the spring is, as in the lattice without pressure,

$$f_n = - d\phi(r_n)/dr_n = a(e^{-br_n} - 1) \tag{J.9}$$

which can be written as

$$f_n = a'(e^{-b(r_n+\sigma')} - 1) + f_{\mathrm{ext}} \ . \tag{J.10}$$

Since the time average of (J.7) gives

$$\overline{e^{-b(r_n+\sigma')} - 1} = 0 \ , \qquad \text{we have} \tag{J.11}$$

$$\bar{f}_n = f_{\mathrm{ext}} \ . \tag{J.12}$$

Thus, the average force of the spring is equal to the external pressure, which is a natural result.

II) When a constant pressure f_{ext} is present, we have only to replace r_n by $r_n + \sigma'$ and a by a'. Thus, in general, we have

$$e^{-b(r_n+\sigma')} - 1 = \frac{1}{a'} \frac{d^2}{dt^2} S_n \tag{J.13}$$

$$r_n + \sigma' = \frac{1}{m} (2S_n - S_{n-1} - S_{n+1}) \tag{J.14}$$

and the equations of motion are

$$\ln \left(1 + \frac{1}{a'} \frac{d^2}{dt^2} S_n\right) = \frac{b}{m} (S_{n-1} + S_{n+1} - 2S_n) \ . \tag{J.15}$$

Equation (J.14) is obtained by integrating $m\ddot{y}_n = s_{n-1} - s_n$. It is to be understood that S_n includes an additional term which is independent of time, but may depend on n.

III) We notice the following formula for the ϑ_0 functions:

$$\frac{\vartheta_0(x + y)\, \vartheta_0(x - y)}{[\vartheta_0(x)]^2} = C \left[1 + v_1^2 \frac{d^2}{dx^2} \ln \vartheta_0(x)\right] \qquad \text{with} \tag{J.16}$$

$$C = C(y) = \left[\frac{\vartheta_0(y)}{\vartheta_0(0)}\right]^2 \left[1 - \left(1 - \frac{E}{K}\right) \mathrm{sn}^2 (2Ky)\right] \tag{J.17}$$

$$v_1^2 = v_1^2 (y) = \mathrm{sn}^2 (2Ky) \Big/ (2K)^2 \left[1 - \left(1 - \frac{E}{K}\right) \mathrm{sn}^2 (2Ky)\right] \ . \tag{J.18}$$

Putting $y = 1/\lambda$, we obtain a solution

$$S_n = \frac{m}{b} \ln\left[\vartheta_0\left(\nu t - \frac{n}{\lambda}\right)\Big/ C^{n^2/2}\right] \tag{J.19}$$

to the equations of motion (J.15). Then (J.13) gives a cnoidal wave solution (J.7). In order to satisfy the equations of motion, we have the dispersion relation

$$\nu^2 = \frac{a'b}{m}\nu_1^2 = \frac{a'b}{m}\frac{1}{(2K)^2}\frac{\mathrm{sn}^2\,(2K/\lambda)}{1-(1-E/K)\,\mathrm{sn}^2(2K/\lambda)} \tag{J.20}$$

which is, of course, the same as (J.8).

IV) When the lattice is free to expand, the dispersion relation for a cnoidal wave is

$$\omega = \frac{\pi}{K(k)}\sqrt{\frac{ab}{m}}\Big/\sqrt{\frac{\mathrm{cn}^2\,(2K/\lambda)}{\mathrm{sn}^2\,(2K/\lambda)} + \frac{E(k)}{K(k)}} \tag{J.21}$$

and the expansion is

$$\bar{r}_n = \frac{1}{b}\ln C \qquad \text{with} \tag{J.22}$$

$$C = \left(\frac{\vartheta_0(1/\lambda)}{\vartheta_0(0)}\right)^2\left[1 - \left(1 - \frac{E(k)}{K(k)}\right)\mathrm{sn}^2\frac{2K}{\lambda}\right]. \tag{J.23}$$

If the modulus k of the cnoidal wave is increased due to the expansion of the lattice, the frequency does not increase, but decreases. In fact, the frequency ω_m for the smallest wavelength $\lambda = 2$ is

$$\omega_m = \omega(\lambda = 2) = \pi\sqrt{\frac{ab}{m}}\sqrt{\frac{1}{E(k)\,K(k)}} \tag{J.24}$$

which decreases with increasing modulus k.

Suppose we have an external pressure f_{ext}. If the pressure is such that

$$e^{-b\sigma'} = \frac{a}{a + f_{\text{ext}}} = \frac{a}{a'}, \tag{J.25}$$

then the lattice shrinks by the amount σ'. Now, give a cnoidal wave to the lattice, which then expands by the amount $(1/b)\ln C$. Therefore if we choose the external pressure in such a way that

$$\sigma' = \frac{1}{b}\ln C \tag{J.26}$$

holds, then the expansion due to the cnoidal wave is compensated by the contraction due to the pressure. To eliminate the expansion, we have to apply the pressure given by

$$a' = a + f_{ext} = aC.$$ (J.27)

For the minimum wavelength $\lambda = 2$, we have

$$C(\lambda = 2) = \left[\frac{\vartheta_0(1/2)}{\vartheta_0(0)}\right]^2 \frac{E(k)}{K(k)}$$

$$= \frac{1}{(1 - k^2)^{1/2}} \frac{E(k)}{K(k)},$$ (J.28)

where we have used the formula

$$\left[\frac{\vartheta_0(1/2)}{\vartheta_0(0)}\right]^2 = \left[\frac{\vartheta_3(0)}{\vartheta_0(0)}\right]^2 = \frac{1}{k'} = \frac{1}{(1 - k^2)^{1/2}}.$$ (J.29)

Therefore the maximum frequency with pressure to eliminate expansion is given as

$$\omega_m = \omega(\lambda = 2)$$

$$= \pi \sqrt{\frac{a'b}{m}} \sqrt{\frac{1}{E(k)\,K(k)}}$$

$$= \pi \sqrt{\frac{ab}{m}} \sqrt{C} \sqrt{\frac{1}{E(k)\,K(k)}}$$

$$= \frac{\pi\sqrt{ab/m}}{K(k)(1 - k^2)^{1/4}}.$$ (J.30)

If we estimate values for $k \to 0$ and for $k \to 1$, we have $K(h) \to \pi/2$ for $k \to 0$, and $K(k) \to \ln(4/\sqrt{1 - k^2})$ for $k \to 1$, so that we have

$$\omega_m \to \begin{cases} 2\sqrt{\dfrac{ab}{m}} & (k \to 0) \\[2mm] \dfrac{\pi}{K(k)} \sqrt{\dfrac{ab}{m}}\, e^{K/8} \to \infty & (k \to 1). \end{cases}$$ (J.31)

The value of ω_m for $k \to 0$ is that for a linear lattice. We have $\omega_m \to \infty$ for $k \to 1$ when the expansion is eliminated. This is a consequence of the fact that the nonlinear lattice approaches a hard-sphere system when the energy is increased because of the strong repulsion between particles.

Simplified Answers to Main Problems

Problem 1.1.

$$U = \frac{1}{2}(q_1^2 + q_2^2) + q_1^2 q_2 - \frac{1}{3} q_2^3$$

takes minimum values at $q_1 = q_2 = 0$. Using polar coordinates (r, θ), we write

$$q_1 = r \cos \theta, \qquad q_2 = r \sin \theta$$

to have

$$U = \frac{1}{2} r^2 + \frac{r^3}{3}(3 \cos^2\theta - \sin^2\theta) \sin \theta$$

$$= \frac{1}{2} r^2 + \frac{r^3}{3}(3 - 4 \sin^2\theta) \sin \theta$$

$$= \frac{1}{2} r^2 + \frac{r^3}{3} \sin 3\theta .$$

Thus U has the symmetry of a regular triangle. Further, $\partial U/\partial q_1 = 0$ at $q_2 = -1/4$, and $\partial U/\partial q_1 = \partial U/\partial q_2 = 0$ at $q_1 = 0$, $q_2 = 1$.

Problem 1.2. Write the Hamiltonian for a linear lattice as

$$\mathcal{H} = \sum_{n=1}^{N} \frac{1}{2m_n} P_n^2 + \sum_{n=0}^{N-1} \frac{\kappa_n}{2} (y_{n+1} - y_n)^2$$

$(y_0 = 0)$. By the dual transformation $(y, p) \rightarrow (r, s)$ we have

$$\mathcal{H} = \sum_{n=1}^{N} \frac{1}{2m_n} (s_n - s_{n-1})^2 + \sum_{n=0}^{N-1} \frac{\kappa_n}{2} r_n^2 .$$

Using a canonical transformation

$$r_n = P_{n+1}/a, \qquad s_n = -aQ_{n+1} \qquad (a = \text{const.})$$

we have

$$\mathcal{H} = \sum_{n=1}^{N} \frac{\kappa_n}{2a^2} P_n^2 + \sum_{n=0}^{N-1} \frac{a^2}{2m_n} (Q_{n+1} - Q_n)^2$$

$$= \sum_{n=1}^{N} \frac{1}{2m_n'} P_n^2 + \sum_{n=0}^{N-1} \frac{\kappa_n'}{2} (Q_{n+1} - Q_n)^2$$

with

$$\frac{1}{m_n'} = \frac{\kappa_n}{a^2}, \qquad \kappa_n' = \frac{a^2}{m_n} .$$

Problem 1.3. For $\phi(r) = \alpha(\beta + r^2)^{1/2}$ we have

$$\dot{s} = -\phi'(r) = -\alpha r(\beta + r^2)^{-1/2}$$

which can be written as

$$(\alpha^2 - \dot{s}^2) r^2 = \beta \dot{s}^2 .$$

Thus (considering $\dot{s} \to -\alpha r \beta^{-1/2}$ for $r \to 0$), we have

$$r = \frac{-\sqrt{\beta} \, \dot{s}}{\sqrt{\alpha^2 - \dot{s}^2}} \equiv -\frac{1}{m} \chi(\dot{s}_n)$$

and the equations of motion (1.4.13) reduce to

$$\frac{d}{dt} \frac{m \dot{s}_n}{\sqrt{1 - \dot{s}_n^2/c^2}} = -\gamma(2s_n - s_{n-1} - s_{n+1})$$

with

$$c = \alpha, \qquad \gamma = \alpha/\sqrt{\beta} .$$

In the above equation $m\dot{s}/\sqrt{1 - \dot{s}^2/c^2}$ can be interpreted as the momentum of the special relativity where $\dot{s}$ is the velocity and c is the light speed.

Problem 2.4. Since $\text{sn}K = 1$, for $\lambda = 2$. the dispersion relation (2.3.2) yields

$$(2K\nu)^2/(ab/m) = K/E .$$

Therefore writing $\beta = 2\nu K$, we have

$$e^{-br_n} = \frac{(2K\nu)^2}{ab/m} \, \text{dn}^2 \, [(n \pm 2\nu t)K]$$

$$= \begin{cases} \dfrac{K}{E} \, \text{dn}^2 \beta t & (n = \text{even}) \\[2ex] \dfrac{K}{E} \, k'^2/\text{dn}^2 \beta t & (n = \text{odd}) . \end{cases}$$

Thus, if we write $r_n = \Delta + 2x$ ($n =$ even), and $r_n = \Delta - 2x$ ($n =$ odd), we have

$$e^{-2b\Delta} = \left(\frac{K}{E}\right)^2 k'^2 \qquad \therefore \quad e^{-b\Delta} = \frac{K}{E} k' .$$

Therefore

$$e^{-2bx} = \frac{1}{k'} \, \mathrm{dn}^2 \beta t$$

and the equations of motion (2.2.9) reduce to $m\ddot{x} = -2a'\sinh 2bx$.

Problem 2.5. By definition $\vartheta_0(x) = \vartheta_0(x, q) = \phi(q) \prod_{r=1}^{\infty} (1 - 2q^{2r-1}\cos 2\pi x + q^{4r-2})$ with

$$\phi(q) = \prod_{r=1}^{\infty} (1 - q^{2r}) .$$

q is related to the modulus k by

$$q = e^{-\pi K(k')/K(k)} = e^{-\pi K'/K}$$

where $k' = \sqrt{1 - k^2}$. Noting that

$$(1 - 2q_1 \cos 2\pi x + q_1^2) [1 - 2q_1 \cos 2\pi(x \pm 1/2) + q_1^2]$$
$$= 1 - 2q_1^2 \cos 4\pi x + q_1^4 ,$$

we have

$$\vartheta_0(2x, q)^2 = \varphi(q) \, \vartheta_0(x, q) \, \vartheta_0(x \pm 1/2, q)$$

where $\varphi(q)$ is a certain function of q. Therefore

$$S_n \left(vt - \frac{n}{\lambda}\right) = \ln \vartheta_0 \left(vt - \frac{n}{\lambda}, q\right) + \mathrm{const.}$$

$$= \ln \vartheta_0 \left(\frac{v}{2} t - \frac{n}{2\lambda}, \sqrt{q}\right) + \ln \vartheta_0 \left(\frac{v}{2} t - \frac{n+\lambda}{2\lambda}, \sqrt{q}\right)$$
$$+ \mathrm{const.}$$

$$\equiv S_n' \left(\frac{v}{2} t - \frac{n}{2\lambda}\right) + S_n'\left(\frac{v}{2} t - \frac{n+\lambda}{2\lambda}\right) + \mathrm{const.} ,$$

where S_n' is the value of S_n for a cnoidal wave with a wavelength twice as large, and its parameter is

$$q' \equiv \sqrt{q} = e^{-\pi K(k')/2K(k)} .$$

Therefore, if we introduce a new modulus κ by

$$q' = e^{-\pi K(\kappa')/K(\kappa)} \qquad (\kappa' = \sqrt{1 - \kappa^2}),$$

we have

$$\frac{K(k')}{2K(k)} = \frac{K(\kappa')}{K(\kappa)},$$

where the functions on both sides are monotonously decreasing functions of k and κ, respectively, the left-hand side is ∞ for $k = 0$ and 0 for $k = 1$, and the right-hand side behaves similarly. Thus for $0 < k < 1$, κ is determined uniquely $(0 < \kappa < 1)$. If we write the cnoidal wave by $e^{-r_n} - 1 = d^2 S_n/dt^2$, we have

$$e^{-r_n} - 1 = [2K(k)\nu]^2 \left\{ \mathrm{dn}^2\left[2\left(\nu t - \frac{n}{\lambda}\right) K(k), k\right] - \frac{E(k)}{K(k)} \right\}$$

$$= [K(\kappa)\,\nu]^2 \left\{ \mathrm{dn}^2\left[\left(\nu t - \frac{n}{\lambda}\right) K(\kappa), \kappa\right] - \frac{E(\kappa)}{K(\kappa)} \right\}$$

$$+ [K(\kappa)\,\nu]^2 \left\{ \mathrm{dn}^2\left[\left(\nu t - \frac{n}{\lambda} - 1\right) K(\kappa), \kappa\right] - \frac{E(\kappa)}{K(\kappa)} \right\}.$$

Problem 2.6. Using the hint, from $e^{-r_n} - 1 = d^2 \ln \vartheta_0/dt^2$, we have

$$e^{-r_n} - 1 = [2K(k)\nu]^2 \left\{ \mathrm{dn}^2\left[2\left(\nu t - \frac{n}{\lambda}\right) K(k), k\right] - \frac{E(k)}{K(k)} \right\}$$

$$= \left[2K(\kappa)\frac{\nu}{l}\right]^2 \sum_{s=0}^{l-1} \left\{ \mathrm{dn}^2\left[2\left(\nu t - \frac{n}{l\lambda} + \frac{s}{l}\right) K(\kappa), \kappa\right) - \frac{E(\kappa)}{K(\kappa)} \right\}$$

where κ is related to k by

$$\frac{K(k)}{lK(k)} = \frac{K(\kappa')}{K(\kappa)}.$$

Problem 2.7. In the preceding problem, summation over s may extend from $s = -l/2$ to $(l/2) - 1$. Shifting thus, we let $l \to \infty$, then $\kappa \to 1$, and a cnoidal wave with the wavelength λ is expressed as a sum of solitons with each center at $n = s\lambda$ ($s = -\infty, \ldots, 0, 1, \ldots, \infty$). The result is (2.3.8).

Problem 2.8. For simplicity we use dimensionless formulas. Then

$$e^{-(Q_{n+1} - Q_n)} = \beta^2 \mathrm{sech}^2 (\kappa n - \beta t)$$

$$s_n = -\beta \tanh (\kappa n - \beta t) + \mathrm{const.}$$

$$S_n = \ln \cosh (\kappa n - \beta t) + (\text{linear term with respect to } n \text{ and } t)$$

$$Q_n = S_{n-1} - S_n = \ln \frac{\cosh [\kappa(n - 1) - \beta t]}{\cosh (\kappa n - \beta t)} + \mathrm{const.},$$

where

$$\beta = \sinh \kappa .$$

Therefore

$$P_n = \dot{Q}_n = s_{n-1} - s_n$$
$$= \beta \{\tanh [\beta t - \kappa(n - 1)] - \tanh (\beta t - \kappa n)\}$$

and

$$M = Q_{-\infty} - Q_\infty = 2\kappa$$

$$P = \sum_{n=-\infty}^{\infty} \dot{Q}_n = s_{-\infty} - s_\infty = 2\beta = Mc ,$$

where $c = \beta/\kappa$ is the speed of the soliton.

Problem 2.9. We restore the dimensions in the preceding formulas

$$E = \sum_n \frac{a}{b} [e^{-(Q_{n+1}-Q_n)} - 1 + (Q_{n+1} - Q_n)] + \sum_n \frac{a}{2b} \dot{Q}_n^2$$

$$= \sum_n \frac{a}{b} [\beta^2 \operatorname{sech}^2 (\kappa n - \beta t) + (Q_{n+1} - Q_n)]$$

$$+ \sum_n \frac{a}{2b} \beta^2 \{\tanh [\beta t - \kappa(n - 1)] - \tanh (\beta t - \kappa n)\}^2$$

$$= \frac{a}{b} \beta^2 \sum_n [1 - \tanh^2 (\kappa n - \beta t)] - \frac{2a\kappa}{b}$$

$$+ \frac{a}{2b} \beta^2 \sum_n \{\tanh^2 [\kappa(n - 1) - \beta t] + \tanh^2(\kappa n - \beta t)\}$$

$$- 2 \tanh [\kappa(n - 1) - \beta t] \tanh (\kappa n - \beta t)\}$$

$$= -\frac{2a\kappa}{b} + \frac{a}{b} \beta^2 \sum_n \frac{1}{2} \{\tanh^2 [\kappa(n - 1) - \beta t] - \tanh^2 (\kappa n - \beta t)\}$$

$$+ \frac{a}{b} \beta^2 \sum_n \frac{\tanh [\kappa(n - 1) - \beta t] - \tanh (\kappa n - \beta t)}{\tanh \kappa} ,$$

where we have used the identity

$$1 - \tanh \alpha \tanh \beta = \frac{\tanh \alpha - \tanh \beta}{\tanh (\alpha - \beta)} .$$

Further, using $\tanh(\kappa n - \beta t) \to \pm 1$ for $n \to \pm \infty$, we have

$$E = -\frac{2a\kappa}{b} + \frac{2a}{b} \frac{\beta^2}{\tanh \kappa}$$

$$= \frac{2a}{b} (\sinh \kappa \cosh \kappa - \kappa) .$$

Problem 3.1. We let $n \rightarrow -\infty$ in (3.6.7), with $|z_1| < 1$, to get

$$c_1 A^{(n)} \rightarrow -c_1(t)^2 e^{-2\delta} z_1^{-n-2} = -(1 - z_1^2) z_1^{-n-2}$$

where (3.6.8) is used. Further, since $K(n, n) \rightarrow z_1$ for $n \rightarrow -\infty$ [cf. (3.6.10)],

$$\phi(n, z) \rightarrow z_1 z^n \left[1 - (1 - z_1^2) z_1^{-n-2} \frac{z_1^{n+1}}{z^{-1} - z_1} \right]$$

$$= z_1 z^n \left(1 - \frac{z_1^{-1} - z_1}{z^{-1} - z_1} \right)$$

$$= z^n z_1 \frac{z^{-1} - z_1^{-1}}{z^{-1} - z_1} .$$

If we replace z by z^{-1}, we have

$$\phi(n, z^{-1}) \rightarrow z^{-n} \frac{1 - zz_1}{z_1 - z} .$$

Thus

$$e^{-i\eta} = \frac{1 - zz_1}{z_1 - z} .$$

On the other hand, from the definition of the transmission coefficient, when $R(z, t) = \beta(z, t) = 0$, we have

$$S(n, z, t) = \frac{\psi(n, z, t)}{\alpha(z)} = \phi(n, z^{-1}) e^{i\omega t} .$$

Since $\psi(n, z, t) \rightarrow z^{-n} e^{i\omega t}$ for $n \rightarrow -\infty$, we have $1/\alpha(z) = e^{-i\eta}$.

Problem 4.2. In (4.4.1), we put

$$\omega = df = \omega_l, \qquad \eta = \omega_k .$$

Then we have

$$A_i = \int_{\alpha_i} \omega_l = \delta_{il}, \qquad B_i = \int_{\beta_i} \omega_l = \tau_{il}$$

$$A_i' = \int_{\alpha_i} \omega_k = \delta_{ik}, \qquad B_i' = \int_{\beta_i} \omega_k = \tau_{ik} .$$

ω_l and ω_k have no pole (Abelian differentials of the first kind), and f and η are differentials with no poles. Therefore, by (4.4.7) we have

$$0 = \sum_j (\delta_{jl}\tau_{jk} - \tau_{jl}\delta_{jk}) = \tau_{lk} - \tau_{kl} .$$

References

Chapter 1

1.1 (a) E. Fermi, J. Pasta, S. Ulam: Los Alamos Rpt. LA-1940 (1955) *Collected Papers of Enrico Fermi* (Univ. of Chicago Press, Chicago 1965) Vol. II, p. 978
   (b) J. L. Tuck: Los Alamos Rpt. LA-3990 (1968): J. L. Tuck, M. T. Menzel: Adv. Math. **9**, 339 – 407 (1972)
   (c) N. J. Zabusky: In *Mathematical Models in Physical Sciences*, ed. by S. Drobot (Prentice-Hall, Englewood Cliffs, NJ 1963)
   (d) R. S. Northcote, R. B. Potts: J. Math. Phys. **5**, 85 (1964)
1.2 (a) J. Ford: J. Math. Phys. **2**, 387 (1961)
   (b) J. Ford, J. Waters: J. Math. Phys. **4**, 1293 (1964)
   (c) J. Waters, J. Ford: J. Math. Phys. **7**, 399 (1966)
1.3 (a) E. A. Jackson: J. Math. Phys. **4**, 551 (1963)
   (b) K. Miura: Thesis, Dept. of Computer Science, University of Illinois (1973)
1.4 (a) N. Saito, H. Hirooka: J. Phys. Soc. Jpn. **23**, 167 (1967)
   (b) H. Hirooka, N. Saito: J. Phys. Soc. Jpn. **26**, 624 (1969)
   (c) N. Ooyama, H. Hirooka, N. Saito: J. Phys. Soc. Jpn. **27**, 815 (1969)
   (d) N. Saito, N. Ooyama, Y. Aizawa, H. Hirooka: Prog. Theor. Phys. Suppl. **45**, 209 (1970)
   (e) R. I. Bivins, N. Metropolis, J. R. Pasta: J. Comp. Phys. **12**, 65 (1973)
   (f) N. J. Zabusky: Comput. Phys. Commun. **5**, 1 (1973)
1.5 (a) N. J. Zabusky, G. S. Deem: J. Comp. Phys. **2**, 207 (1968)
   (b) N. J. Zabusky: J. Phys. Soc. Jpn. Suppl. **26**, 196 (1969)
   (c) R. D. Tappert, C. N. Judice: Phys. Rev. Lett. **29**, 1308 (1972)
1.6 (a) M. Toda: Proc. Int. Conf. Statistical Mechanics, Kyoto, 1968; J. Phys. Soc. Jpn. Suppl. **26**, 235 (1969)
   (b) M. Toda: Prog. Theor. Phys. Suppl. **45**, 174 (1970)
1.7 (a) F. M. Israiliev, B. V. Chirikov: Sov. Phys. Dokl. **11**, 30 (1966)
   (b) G. M. Zaslavsky, R. Z. Sagdeev: Sov. Phys. JETP **25**, 718 (1967)
1.8 G. H. Lunsford, J. Ford: J. Math. Phys. **13**, 700 (1972)
1.9 H. Hénon, C. Heiles: Astron. J. **69**, 73 (1964)
1.10 (a) G. Contropoulos: Astron. J. **76**, 147 (1971)
   (b) G. H. Walker, J. Ford: Phys. Rev. **188**, 416 (1969)
1.11 (a) M. Toda: Phys. Lett. A **48**, 335 (1974)
   (b) G. Casati: Lett. Nuovo Cimento **14**, 311 (1975)
   (c) G. Benettin, R. Brambilla, L. Galgani: Physica A **87**, 381 (1977)
1.12 (a) A. S. Wightman: "Statistical Mechanics and Ergodic Theory", in *Statistical Mechanics at the Turn of the Decade*, ed. by E. G. D. Cohen (Dekker, New York 1971)
   (b) V. I. Arnod, A. Avez, *Problèms ergodiques de la mécanique classique* (Gauthier-Villars, Paris 1967); *Ergodic Problems of Classical Mechanics* (Benjamin, New York 1968)
   (c) J. Ford: Adv. Chem. Phys. **24**, 155 (1973)
1.13 D. J. Korteweg, G. de Vries: Philos. Mag. **39**, 422 (1895)
1.14 N. J. Zabusky, M. D. Kruskal: Phys. Rev. Lett. **15**, 240 (1965)
1.15 (a) D. N. Payton, R. Rich, W. M. Visscher: Phys. Rev. **160**, 129 (1967); *Proc. Int. Conf. on Localized Excitations in Solids*, California 1967 (Plenum, New York) p. 657

(b) M. Rich, W. M. Visscher, D. N. Payton: Phys. Rev. A**4**, 1682 (1971)
(c) E. A. Jackson, J. R. Pasta, J. F. Waters: J. Comp. Phys. **2**, 1207 (1968)
1.16 M. Toda: J. Phys. Soc. Jpn. **20**, 2095 (1965)
1.17 M. Toda: Prog. Theor. Phys. Suppl. **36**, 113 (1966)

Chapter 2

2.1 (a) M. Toda: J. Phys. Soc. Jpn. **22**, 431 (1967)
    (b) G. B. Whitham: *Linear and Nonlinear Waves* (Wiley, New York 1973) p. 612
2.2 M. Toda: J. Phys. Soc. Jpn. **23**, 501 (1967)
2.3 cf. [1.6a]
2.4 (a) cf. [1.6b]
    (b) M. Toda, M. Wadati: J. Phys. Soc. Jpn. **34**, 18 (1973)
2.5 (a) N. Saito, N. Ooyama, Y. Aizawa: RIMS Rpt. Kyoto Univ. **171**, 191 (1973)
    (b) N. Ooyama, N. Saito: Prog. Phys. Suppl. **45**, 201 (1970)
2.6 R. Hirota: J. Phys. Soc. Jpn. **35**, 286 (1973)
2.7 R. Hirota, K. Suzuki: J. Phys. Soc. Jpn. **28**, 1388 (1970); Proc. IEEE **51**, 1483 (1973)
2.8 R. Hirota, J. Satsuma: Prog. Theor. Phys. Suppl. **59**, 64 (1976)
2.9 A. C. Scott. F. Y. F. Chu, D. W. Mclaughlin: Proc. IEEE **61**, 1443 (1973)
2.10 (a) J. Ford: Paper Presented at the Int. Conf. on Point Mappings and Their Applications, Toulouse, Sept. 1973
     (b) J. Ford, S. D. Stoddard, J. S. Turner: Prog. Theor. Phys. **50**, 1547 (1973)
2.11 M. Hénon: Phys. Rev. B**9**, 1921 (1974)
2.12 K. Sawada, T. Kotera: Prog. Theor. Phys. Suppl. **59**, 101 (1976)

Chapter 3

3.1 H. Flaschka: Phys. Rev. B**9**, 1924 (1974)
3.2 P. D. Lax: Comm. Pure and Appl. Math. **21**, 467 (1968)
3.3 (a) K. M. Case, M. Kac: J. Math. Phys. **14**, 594 (1973)
    (b) K. M. Case: J. Math. Phys. **14**, 916 (1973)
3.4 H. Flaschka: Prog. Theor. Phys. **51**, 703 (1974)
3.5 (a) I. Kay, H. E. Moses: J. Appl. Phys. **27**, 1503 (1956)
    (b) M. Toda: Phys. Norv. **5**, 203 (1971)
    (c) G. Papini: Bull. Math. Biol. **39**, 129 (1977)
3.6 M. Toda: Phys. Rep. **18C**, 1 (1974): Ark. Fys. Semin. Trondheim 2 (1974)
3.7 J. Moser (ed): *Dynamical Systems, Theory and Application*, Lecture Notes in Physics, Vol. 38 (Springer, Berlin, Heidelberg, New York 1975) p. 467; Adv. Math. **16**, 197 (1975)
3.8 M. Kac, P. van Moerbeke: Adv. in Math. **16**, 160 (1975)
3.9 M. Toda, M. Wadati: J. Phys. Soc. **39**, 1204 (1975)
3.10 M. Wadati, M. Toda: J. Phys. Soc. **39**, 196 (1975)
3.11 H. Chen, C. Liu: J. Math. Phys. **16**, 1428 (1975)
3.12 T. J. Stieltijes: Faculté de Toulouse VIII, 1 (1894)
3.13 K. Sawada, T. Kotera: J. Phys. Soc. Jpn. **44**, 655 (1977); cf. [2.12]
3.14 T. Kotera, S. Yamazaki: J. Phys. Soc. Jpn. **43**, 1797 (1977)
3.15 (a) cf. [2.2]
     (b) cf. [3.5b]
     (c) M. Toda: *Mathematical Problems in Physics*, Lecture Notes in Physics, Vol. 39 (Springer, Berlin, Heidelberg, New York 1975) p. 387
3.16 M. Toda: Prog. Theor. Phys. Suppl. **59**, 1 (1976)
3.17 (a) I. M. Gel'fand, B. M. Levitan: Amer. Math. Soc. Transl. Ser. 2, **1**, 253 (1955)
     (b) I. Kay, H. E. Moses: Nuovo Cimento **3**, 276 (1956)
     (c) I. Kay, H. E. Moses: J. Appl. Phys. **27**, 1503 (1956)

(d) Z. S. Agranovich, V. A. Marchenko: *The Inverse Problem of Scattering Theory* (Gordon and Breach, London 1963) p. 22

(e) L. D. Faddeev: Amer. Math. Soc. Transl. Ser. 2, **65**, 139 (1967)

3.18 M. Wadati, H. Sanuki, K. Konno: Prog. Theor. Phys. **53**, 419 (1975)

Chapter 4

4.1 cf. [3.3]

4.2 (a) M. Kac, P. van Moerbeke: Proc. Nat. Acad. Sci. USA **72**, 1627, 2879 (1975)

(b) P. van Moerbeke: Invent. Math. **37**, 45 (1976)

4.3 E. Date, S. Tanaka: Prog. Theor. Phys. **55**, 457 (1976); Prog. Theor. Phys. Suppl. **59**, 107 (1976)

4.4 (a) C. L. Siegel: *Topics in Complex Function Theory I, II, III*, (Wiley-Interscience, New York 1971)

(b) G. Springer: *Introduction to Riemann Surfaces* (Addison-Wesley, Reading, MA 1957)

(c) Y. Kusunoki: *Theory of Functions* (Asakura, Tokyo 1973) [in Japanese]

4.5 Elliptic Functions:

(a) E. T. Whittaker, G. N. Watson: *A Course of Modern Analysis*, 4th ed. (Cambridge Univ. Press, Cambridge 1946)

(b) A. G. Greenhill: *The Application of Elliptic Functions* (Dover, New York 1957)

(c) H. Jeffreys, B. S. Jeffreys: *Mathematical Physics*, 3rd ed. (Cambridge Univ. Press, Cambridge 1956) Chap. 25

(d) H. Hancoch: *Theory of Elliptic Functions* (Dover, New York 1958)

(e) F. Tricomi: *Elliptische Funktionen* (Akademische Verlagsgesellschaft, Leipzig 1948)

(f) S. Tomochika: *Theory of Elliptic Functions* (Kyoritsu, Tokyo 1958) [in Japanese]

(g) S. Ando: *Introduction to Elliptic Integrals and Elliptic Functions* (Nisshin. Tokyo 1970) [in Japanese]

(h) M. Toda: *Introduction to Elliptic Functions* (Nihon-Hyoron, Tokyo 1976) [in Japanese]

4.6 Mathieu Equation and Hill's Equation:

(a) E. T. Whittaker, G. N. Watson: *A Course of Modern Analysis*, 4th ed. (Cambridge Univ. Press, Cambridge 1946)

(b) H. P. McKean, P. van Moerbeke: Invent Math. **30**, 217 (1974)

(c) H. Hochstadt: *The Functions of Mathematical Physics* (Wiley, New York 1971)

(d) K. Husimi: *Classical Mechanics* (Iwanami, Tokyo 1964) p. 287 [in Japanese]

4.7 M. Toda: Ark. Fys. Semin. Trondheim 3 (1977)

4.8 Russian literature on the analysis under periodic boundary conditions are cited in [4.3].
Further, I. M. Krichever, Dokl. Akad. Nauk USSR **227**, 291 (1976); Funct. Anal. Appl. **10**, 75 (1976)

Chapter 5

5.1 (a) McLaughlin: J. Math. Phys. **16**, 96, 1704 (1975)

(b) H. Flaschka, D. W. McLaughlin: In *Bäcklund Transformations, the Inverse Scattering Method, Solitons, and Their Applications*, ed. by R. M. Miura. Lecture Notes in Mathematics, Vol. 515 (Springer, Berlin, Heidelberg, New York 1976) p. 253

5.2 (a) H. Flaschka, D. W. McLaughlin: Prog. Theor. Phys. **55**, 438 (1976)

(b) H. Flaschka: In [3.7] p. 441

5.3 cf. [4.6b)

Chapter 6

6.1 N. Saitoh: J. Phys. Soc. Jpn. **49**, 409 (1980); **54**, 3261 (1985)
6.2 R. Hirota: In *Nonlinear Integrable Systems − Classical Theory and Quantum Theory*, ed. by M. Jimbo, T. Miwa (World Scientific, Singapore, 1983) p. 17
6.3 T. Yoneyama: J. Phys. Soc. Jpn. **55**, 753 (1986); Prog. Theor. Phys. **72**, 1081 (1984)
6.4 S. Kovalevskaja: Acta. Math. **12**, 177 (1889); **14**, 81 (1890)
6.5 M. J. Ablowitz, A. Ramani, H. Segur: Lett. Nuovo Cimento **23**, 333 (1978); J. Math. Phys. **21**, 715 (1980)
6.6 J. Weiss: J. Math. Phys. **24**, 1405 (1983); **25**, 13 (1984)
6.7 J. D. Gibbon, M. Tabor: J. Math. Phys. **26**, 1956 (1985)
6.8 A. Ramani: private communication
6.9 O. I. Bogoyavlensky: Commun. Math. Phys. **51**, 201 (1976)
6.10 M. A. Olshanetsky, A. M. Perelomov: Inventiones Math. **54**, 261 (1979)
6.11 B. Konstant: Adv. Math. **34**, 195 (1979)
6.12 R. S. Farwell, M. Minami: Prog. Theor. Phys. **69**, 1091 (1983); **70**, 710 (1983)
6.13 H. Yoshida: In *Nonlinear Integrable Systems − Classical Theory and Quantum Theory*, ed. by M. Jimbo, T. Miwa (World Scientific, Singapore, 1983) p. 273
6.14 (a) A. V. Mikhailov: JETP Lett. **30**, 414 (1979)
      (b) G. Darboux: *Leçons sur la théorie générale des surfaces II* (Celsea, New York 1972)
      (c) J. D. Gibbon, M. Tabor: J. Math. Phys. **26**, 1956 (1985)
      (d) Y. Kametaka: Proc. Jpn. Acad. **60A**, 79, 121, 145, 181 (1984)
6.15 S. A. Bulgadaev: Phys. Lett. **96B**, 151 (1980); R. S. Farwell, M. Minami: J. Phys. **A15**, 25 (1982)
6.16 A. Nakamura: J. Phys. Soc. Jpn. **52**, 380 (1983)
6.17 N. Saitoh, S. Takeno, É. I. Takizawa: J. Phys. Soc. Jpn. **54**, 3701 (1985); N. Saitoh, É. I. Takizawa, S. Takeno: J. Phys. Soc. Jpn. **54**, 4524 (1985); N. Saitoh, É. I. Takizawa: J. Phys. Soc. Jpn. **55**, 1827 (1986)
6.18 C. N. Yang, C. P. Yang: J. Math. Phys. **10**, 1115 (1969)
6.19 M. Wadati: J. Phys. Soc. Jpn. **54**, 3327 (1985)
6.20 B. Sutherland: Rocky Mountain J. Math. **8**, 413 (1978)
6.21 M. Opper: Phys. Lett. **112A**, 201 (1985)
6.22 F. G. Mertens: Z. Phys. **B55**, 353 (1984)
6.23 J. Satsuma: In *Nonlinear Integrable Systems − Classical Theory and Quantum Theory*, ed. by M. Jimbo, T. Miwa (World Scientific, Singapore, 1983) p. 183
6.24 M. Toda: unpublished
6.25 E. T. Whittaker, G. N. Watson: *A Course of Modern Analysis* (Cambridge Univ. Press, Cambridge 1973) p. 251
6.26 cf. Y. Ohta, J. Satsuma, D. Takahashi, T. Tokihiro: Prog. Theor. Phys. Suppl. **94**, 210 (1988); E. Date, M. Kashiwara, M. Jimbo, T. Miwa: In *Nonlinear Integrable Systems − Classical Theory and Quantum Theory,* ed. by M. Jimbo, T. Miwa (World Scientific, Singapore, 1983) p. 39
6.27 K. Ueno, K. Takasaki: RIMS-397 (1982), -425, 426 (1983)
6.28 D. N. Payton, M. Rich, W. M. Visscher: Phys. Rev. **160**, 706 (1967); Proc. Int. Conf. on Localized Excitation in Solids, California 1967 (Plenum, New York 1968) p. 657
6.29 M. Toda: Phys. Scripta **20**, 424 (1979)
6.30 S. Watanabe, M. Toda: J. Phys. Soc. Jpn. **50**, 3436 (1981)
6.31 F. Mokross, H. Büttner: J. Phys. **C16**, 4539 (1983)
6.32 M. Toda, R. Hirota, J. Satsuma: Prog. Theor. Phys. Suppl. **59**, 148 (1976)

Appendices

A.1 cf. [3.12]
C.1 cf. [4.6b] H. Hochstadt: Arch. Rat. Math. Anal. **19**, 353 (1965)

F.1   R. Bellman: *A Brief Introduction to Theta Functions* (Holt, Rinehart, and Winston, New York 1961)

G.1   R. Hirota, J. Satsuma: Prog. Theor. Phys. Suppl. **59**, 64 (1976)

G.2   J. H. H. Perk: Phys. Lett. **A79**, 3 (1980);
      H. Au-Yang, J. H. H. Perk: Physica **144**, 44 (1987)

G.3   R. Hirota: J. Phys. Soc. Jpn. **50**, 3785 (1981); In *Nonlinear Integrable Systems − Classical Theory and Quantum Theory,* ed. by M. Jimbo, T. Miwa (World Scientific, Singapore, 1983) p. 15;
      T. Miwa: Proc. Japan Acad. **58A**, 9 (1982)

H.1   (a) N. Saito, N. Hirotomi, A. Ichimura: J. Phys. Soc. Jpn. **39**, 1431 (1975)
      (b) N. Saito, A. Ichimura: Prog. Theor. Phys. Suppl. **59**, 137 (1976)

H.2   R. L. Bivins, N. Metropolis, J. Pasta: J. Comp. Phys. **12**, 65 (1973)

H.3   cf. [1.7]

H.4   N. Saito, N. Ooyama, Y. Aizawa, H. Hirooka: Prog. Theor. Phys. Suppl. **45**, 209 (1970)

I.1   cf. [1.6b]

I.2   H. Takahashi: Proc. Phys. Math. Soc. **24**, 60 (1942); E. H. Lieb, D. C. Mattis: *Mathematical Physics in One Dimension* (Academic, New York 1966) p. 25

Bibliography

The following references have been added in order to bring the reference lists up to date.

General expositions of the soliton theory, Bäcklund transformation and the inverse scattering method can be seen in

R. K. Bullough, P. J. Caudrey (eds.): *Solitons*, Topics in Current Physics, Vol. 17 (Springer, Berlin, Heiderberg, New York 1980),

which includes the following articles referring to the lattice problem:

M. Toda: On a Nonlinear Lattice, pp. 143–155;

R. Hirota: Direct Methods in Soliton Theory, pp. 157–176;

M. Wadati: Generalized Matrix Form of the Inverse Scattering Method, pp. 287–300;

L. D. Faddeev: A Hamiltonian Interpretation of the Inverse Scattering Method, pp. 339–354.

An elaborate mathematical exposition in terms of Lie algebra:

Bertram Kostant: The Solution to a Generalized Toda Lattice and Representation Theory. Adv. Math. **34**, 195–338 (1979).

See also

O. I. Bogoyavlensky: On Perturbation of the Periodic Toda Lattice. Commun. Math. Phys. **51**, 201 (1976)

W. W. Symes: Hamiltonian Group Actions and Integrable Systems. Physica D, **1D**, 339–374 (1980)

Section 1.3. On the thermal conductivity of solids, see also

M. Toda: Solitons and Heat Conduction. Phys. Scr. **20**, 424–430 (1979).

Sections 2.1–2.5. As general review works on the lattice problem, we may refer to

M. Toda: Problems in Nonlinear Dynamics. Rocky Mountain Math. **8**, 197–209 (1978).

Sections 2.7 and 3.11. It was shown that the exponential lattice and the Korteweg-de Vries equation can be entirely connected through integrable regime:

N. Saitoh: A Transformation Connecting the Toda Lattice and the KdV Equation. J. Phys. Soc. Jpn. **49**, 409–416 (1980).

Section 3.8. Certain generalization of the integrable lattice system is given, for example, by

A. V. Mikhailov: Integrability of a Two-Dimensional Generalization of the Toda Chain. JETP Lett. **30**, 414–418 (1979).

Section 4.4. Hyperelliptic curves, or the Riemann surfaces, related to periodic systems are discussed in

H. P. McKean, P. van Moerbeke: Hill and Toda Curves. Commun. Pure Appl. Math. **38**, 23–42 (1980).

Appendix G. On Hirota's Method and the discrete-time lattice, see also

R. Hirota: Difference Analogue of the Korteweg-de Vries Equation. J. Phys. Soc. Jpn. **43**, 1424–1433 (1977);
Discrete-Time Toda Equation. J. Phys. Soc. Jpn. **43**, 2074–2078 (1977);
Discrete Sine-Gordon Equation. J. Phys. Soc. Jpn. **43**, 2079–2086 (1977);
Bäcklund Transformation for the Discrete-Time Toda Equation. J. Phys. Soc. Jpn. **45**, 321–332 (1978).

Subject Index

List of Authors Cited in Text